Children's Food

Children's Food

Marketing and innovation

Edited by

GLEN SMITH
Children's Research Unit
London
United Kingdom

BLACKIE ACADEMIC & PROFESSIONAL

An Imprint of Chapman & Hall

London · Weinheim · New York · Tokyo · Melbourne · Madras

**Published by Blackie Academic & Professional, an imprint of
Chapman & Hall, 2–6 Boundary Row, London SE1 8HN, UK**

Chapman & Hall, 2–6 Boundary Row, London SE1 8HN, UK

Chapman & Hall GmbH, Pappelallee 3, 69469 Weinheim, Germany

Chapman & Hall USA, 115 Fifth Avenue, New York
NY 10003, USA

Chapman & Hall Japan, ITP-Japan, Kyowa Building, 3F, 2-2-1
Hirakawacho, Chiyoda-ku, Tokyo 102, Japan

DA Book (Aust.) Pty Ltd, 648 Whitehorse Road, Mitcham 3132, Victoria,
Australia

Chapman & Hall India, R. Seshadri, 32 Second Main Road, CIT East,
Madras 600 035, India

First edition 1997

© 1997 Chapman & Hall

Typeset in 10/12 Times by Cambrian Typesetters, Frimley, Surrey

Printed in Great Britain by St Edmundsbury Press Ltd, Bury St Edmunds,
Suffolk

ISBN 0 7514 0274 5

A catalogue record for this book is available from the British Library

Library of Congress Catalog Card Number 96–84184

∞ Printed on acid-free text paper, manufactured in accordance with ANSI/
NISO Z39.48-1992 (Permanence of Paper)

Contents

9 Children's views on food and nutrition: a pan-European study   152
JEAN-PIERRE PROPONNET, CHAIRMAN,
EUROPEAN FOOD INFORMATION COUNCIL (EUFIC)

Contributors

Professor V.N. Balasubramanyam The Management School, Lancaster University, Department of Economics Lancaster, LA1 4YX, UK

Professor T.P. Barwise London Business School, Sussex Place, Regents Park, London NW1 4SA, UK

Mrs S.H.L. Clark Formerly of PI Design International, 1–5 Colville Mews, Lonsdale Road, London W11 2AR, UK

Mr Jean-Pierre Proponnet EUFIC, 225 Avenue Louise, Box 5, B-1050, Brussels

Dr N.J. Jardine Health & Nutrition Sciences, Nestlé York Ltd., PO Box 204, York YO1 1XY, UK

Mr S. Lang Formerly of Henley Centre, 9 Bridewell Place, Blackfriars, London EC4V 6AY, UK

Ms J. Mathews J. Walter Thompson, 40 Berkeley Square, London W1X 6AD, UK

Dr K.R. O'Sullivan Kellogg Company of Great Britain Limited, The Kellogg Building, Talbot Road, Manchester M16 0PU, UK

Mr L. Stanbrook Advertising Association, Abford House, 15 Wilton Road, London SW1V 1NJ, UK

Professor P. Stratton Department of Psychology, Leeds Family Therapy and Research Centre, University of Leeds, Leeds LS2 9JT, UK

Preface

The purpose of this book is to serve as essential reading for those innovating and marketing food products for children as well as those determined to better understand the children's marketplace in order to ensure that it is administered in a manner consistent with the long-term aspirations of society.

The book begins by setting the scene and looking at the way children influence food choices within the family and the role advertising is thought to play in driving those choices. Professor Stratton of The Psychology Business (Department of Psychology, Leeds University) has world renowned expertise in the methodology of researching family dynamics and he shows which are the prime influences on the family diet. J.W. Thompson Advertising Board Director Jane Mathews then evaluates what constitutes effective advertising and reveals enduring themes within the children's marketplace. In Chapter 3, Dr Kathryn O'Sullivan of the Kellogg's company examines the nutritional importance of food under the title 'Starting the day right'. She demonstrates her expertise for introducing young taste buds to products which 'Break the fast'. Simon Lang, Senior Consultant at the Henley Centre follows by examining not only why food tastes change in children but also why family eating is itself changing and the implications for the future.

In Chapter 5, Dr Nick Jardine and Caryn Philpott from Nestlé look at the current status of children's nutrition, whether there is need for change and, if so, how that may be brought about. Lionel Stanbrook, Director of Political Issues of the Advertising Association then comprehensively examines in detail the politics of advertising to children in Europe in Chapter 6. In Chapter 7, Sheila Clark, formerly Chairperson of PI Design Consultants, examines how packaging works with children.

The theme of advertising is continued in Chapter 8. Professor T.P. Barwise (Director of the Centre of Marketing, London Business School) has taken a critical look at the scientific literature which examines the role of advertising in driving choices and in particular considers the evidence offered when examining the hypothesis that children represent a particularly vulnerable market which needs extraordinary protection from the influences of advertising.

Jean-Pierre Proponnet, Chairman of the European Food Information Council (EUFIC), reveals for the first time comparative data of what children eat and drink (over 24 hours) in the UK, France, Germany and

Italy. In addition, the extent of children's nutrition knowledge is revealed and tested in terms of a number of food, health and hygiene concepts. Finally, Professor Balasubramanyam of Lancaster University (Development Economics) examines the inter-national influences on children's food.

Overall, the authors have contributed a wealth of information of great pertinence to all those concerned with innovation and marketing of food products to today's well informed young consumers.

Glen Smith
Chairman
Children's Research Unit

1 Influences on food choice within the family
P. STRATTON

1.1 Introduction

Any attempt to influence children's choice of food will only succeed if it deals realistically with all of the relevant factors. This chapter first specifies the scope that an understanding of food choice must cover, then presents some of the major findings from our studies in the area. These findings are used as a framework for a broader consideration of the kinds of processes involved. The unique role of food choice, at the centre of biological and social interaction, means that nutrition is influenced by many more things than just availability of nutritious food, and food choice has many implications for other aspects of life. Finally, the concepts developed in the chapter are applied to practical issues of effective advertising and marketing, and the potential for the food industry to use the complexity of food choice to make a positive contribution to family life.

The development and marketing of food and drink products for children involves issues that are not only complex, but also sensitive. Children's growth and health, and many aspects of family functioning may be affected by the kinds of food that are made available, and the ways they are promoted. The central role that food and drink play in our lives, and particularly the lives of our children also means that marketing may be tapping into very powerful processes. To be both ethical and successful, marketing must make constructive use of children's motivations and parental concerns in ways that avoid aggravating any of the real problems that surround the issue of children's eating. In fact, it is hoped that this chapter will show that there is scope for the food industry to make a substantial positive contribution to the welfare of children and families. Before examining the processes of food choice in detail, there is a review of the various ways in which eating is significant, and current perceptions of the major problems in children's eating.

1.2 Why food is so significant

Any choice is a decision that is influenced by the context in which it was made as well as by the many aspects of the person making the decision.

Many food choices are made in the context of the family, and family processes around food are a central consideration of this chapter. In external contexts such as supermarkets, family influences will still operate on the individual making the choices. We therefore start with a brief survey of the kinds of role that food and eating play within the family.

If family food choices were simple rational decisions based on nutrition and palatability, the task of meeting the needs of families would be very simple. But providing food within the family is an emotionally charged issue. It is difficult for parents to cope with having the food that they have prepared seemingly casually rejected, yet this is a trick that most children have learned long before their first birthday. The fact is that issues of what is eaten, how much, when, and in what way, become a matter for negotiation between children and their parents. Meals in whatever form are contexts in which important processes of socialization, communication and imitation can occur. For many families mealtimes are the only occasion in which the whole family is regularly together. However, increasing numbers of families are organizing their eating in ways that make shared meals less and less frequent. Many parents are unhappy about this and resent external influences which they see as contributing to the trend.

Feeding and mealtimes make a major contribution to the development of the child. This contribution often depends upon naturally occurring family interactions which are carried out without any specific awareness or intention towards the function they serve. An idea of how significant children's eating is to the parent, and how socially rich the process can become can be gained from studies of early infancy. In this context it is very easy to see how a powerful biological requirement such as food becomes built into many other aspects of family life. Within a few days of life babies have learned that being held in a certain way means that they are about to be fed and, at the same time, if breast fed, they have learned to prefer the smell of their own mother's milk to that of anyone else. It does not take very long before this early learning becomes used as a significant factor in the baby's relationships. The fundamental biological requirements of getting enough food to grow, have become woven into the whole complex of processes by which attachments are formed between babies and their caregivers (Stratton, 1982).

From detailed observations of family meals, Valsiner (1987) demonstrates how they serve to carry many implicit cultural values – for example a typical high chair is a device to limit, but not eliminate, a child's freedom of movement while allowing the child to have his/her attention focused on food and utensils, be at eye level with adults, but also with scope for entertainment between bouts of eating. This situation can then be exploited to help the child acquire motor skills of grasping, manipulating objects and co-ordinating hand, eye and mouth – all needed when using a spoon to get food from the bowl to the mouth. As the child grows, more complex skills such as restraint and manners may be encouraged. Erikson (1963) describes how the Yoruk Indians trained their children to never take food without asking, to eat slowly and maintain silence so that everyone could think about salmon and becoming rich.

When things are going well a meal is a pleasurable process which facilitates positive interaction. It is an occasion in which providers are seen to provide, and where children can copy adult activities and so feel that they are participating helpfully. The activities at the table provide a structure for multiple small-scale interactions, and the pleasure of eating can facilitate other forms of positive interaction. For example, even conversations about school are expected to be more positive during a meal. Mealtimes can therefore enrich the experience and expand the social repertoire of the people involved.

When things are going badly, a meal can become a battle-ground in which each person uses the power of their situation to attempt to win wars over control, independence, recognition and guilt. Refusing to eat is not the only source of power for a child. If sufficiently distressed they can ensure that their distress is shared by everyone else. They will then find it easier to gain acceptance of a refusal to participate in meals and thereby change the whole pattern of family eating.

Of course, most families find a route between these extremes. Much of our research has been to investigate just how the ways that families negotiate their eating affect, and are influenced by, their response to external influences such as advertising, category availability and packaging.

1.3 Personal factors in children's food choice

While trying to cope with some of the complexity of family processes it is important to remember that each child brings their own unique contribution to the equation. First, there are developmental changes in the balance of taste sensitivities which mean that preferences change over time. To some extent taste preferences are a guide to physiological need. It is well known that in a condition of sodium depletion, there will be a spontaneous increase in desire for salt. For children between 2.5 and 5 years, Birch (1987) found that the strongest factor in food preferences was sweetness. However, the second significant factor was familiarity: children preferred those foods with which they were most familiar. It is of great importance that only one basic biological aspect was stronger than familiarity. While sweetness is intrinsic, familiarity is strongly affected by the actions of parents in choosing which foods should be presented to the child. In fact, even within the age range that Birch studied, the effect of simple exposure was diminishing by the age of five.

Some factors which are known to be important later in childhood involve the ways that parents attempt to encourage children to eat. The research described later in this chapter has shown that the dominant issue for parents is to get their children to eat, but some of their methods are counterproductive. Research reviewed by Birch (1987) has shown that using a reward to persuade a child to eat a certain food ('if you eat your peas you can watch TV') actually reduces the child's liking for that food. Worse, if the reward is a food ('if you eat your peas you can have a sweet'), the type of food functioning as a reward becomes even more preferred. The child will end up more resistant to eating peas and more likely to demand sweets.

It is important not to apply adult judgements to children's food preferences. The liking for sweetness and for fats is appropriate to the phase during which the child must not only sustain a level of energy expenditure (which may be relatively greater than in adult life) but also grow. It seems likely that current attitudes to problems of obesity have made parents and others over-concerned about children's consumption of energy dense snacks because they are likely to be high in unrefined sugars and saturated fats. As Poskitt (1994) points out it is not the consumption of such 'junk' foods itself that is a concern because they are not necessarily bad for children. Rather it is the possibility that they may come to dominate the diet and thereby exclude other foods which carry different nutrients. Parents therefore have a legitimate concern to maintain variety in their children's diets.

1.4   Social factors in children's food choices

Research shows that social factors can be powerful influences on food choice. Birch (1980) showed experimentally that children will change their preferences to fit in with their peers. Harper and Sanders (1975) also showed that children were more likely to try unfamiliar foods if adults ate these foods with the children than if the adults merely offered the foods to them. One of our studies with young children with very poor eating habits found that these children were more likely to be eating on their own whereas good eaters almost always ate in the company of other members of the family. We also found evidence that the underfed children were likely to have all of their eating restricted to three set meal times a day, whereas the control children had snacks between meals.

The history of positive and negative experiences with any specific food will strongly influence current preferences. So will observation of how significant other people treat the food. From an early age, but increasingly as the child grows older, the current rulings by the peer group will have a strong influence on the acceptability of foods. Children are becoming increasingly knowledgeable about the relationships between food and health, energy, body shape, and moral issues such as the use of animals. All such knowledge has the potential to be either positive or negative in its effects on the diet.

In most areas of food choice, influences work in both directions. While social factors influence the choices made, it is also the case that the processes around food are used to socialize the child. Advertising which ignores this fact may encounter strong negative reactions. Families in all cultures use eating as a context in which to establish patterns of behaviour and interaction. Some nice examples come from advice offered to children in the sixteenth century, quoted by Valsiner (1987). These include telling children that they should wash their hands; not offer food to another person after it has been in their own mouth (unless the other person is a servant); if you must pick your teeth be discreet about it; do not fish lice out of your hair; and quarrelling at table is most despicable. Advertising which is judged to undermine the use of mealtimes to foster good social behaviour and positive interactions will be resented. More positively, an awareness of the ways that families use eating for these purposes can be used to enhance the value placed on types of food by parents, and increase the effectiveness of advertising.

Anyone who intends to influence children's diets, whether from a health education or an advertising position will need to fit their intervention into the framework of factors which influence children's eating. As this part of the review has shown, this framework is strongly influenced by the way the family and the peer group operate. Interventions which attempt to change a single aspect of diet are likely to be defeated by the multitude of other factors built into the existing pattern. They are therefore unlikely to be successful unless they incorporate the way the whole social system of the child will respond.

1.5 Concerns about children's eating

Parental worries about their children's eating must be universal, but in recent times concerns about eating have become a major preoccupation of Western nations. Many of these concerns which are directed towards adults seem to have become applied to children. It is worthwhile to separate parental concerns which are acknowledged to sometimes be unrealistic from the concerns of nutritionists which are believed by experts to be realistic.

Broadly we can see an age progression. Before the age of five, the most serious concerns are with children who are underfed. This is primarily an issue of energy intake. With anything like a normal diet it is, for example, very unlikely that a child in this age group will achieve adequate energy intake without being well above the recommended minimum for protein.

The reasons for underfeeding vary, but in one way or another include a simple failure of carers to provide for the nutritional requirements of the infant or young child (Hanks and Hobbs, 1993). This failure remains the adults' responsibility even if the situation reaches a stage in which the child refuses to eat. At an extreme, the growth of the child will be affected, a condition known as 'non-organic failure to thrive'. Short periods of undernutrition, even if severe enough to reduce growth, are likely to be followed by a period of 'catch-up growth' (Tanner, 1978). However, underfeeding is of particular concern at this age because if it becomes built into a consistent and long lasting pattern, the consequences can be permanent.

After the age of five, obesity starts to become a more common concern. It is probable that the change is related to the increasing ability of the child to acquire their own food supplies. Obesity is regarded as 'now a major health hazard in technologically sophisticated countries and this condition has been steadily increasing for the last 40 years' (Blundell and Bauer, 1994, p. 11). However, it is also believed to lead to eating disorders, such as anorexia and bulimia, which these authors go on to say: 'constitute one of the most prevalent and disabling conditions afflicting young adults and children'. From the age of about ten we, therefore, need to be alert to a whole range of disorders in which 'there is excessive concern with the control of body weight and shape, accompanied by grossly inadequate, irregular, or chaotic food intake.' (Bryant-Waugh and Lask, 1995, p. 191). The incidence of eating disorders in childhood is not known but it is worth noting that Bryant-Waugh and Lask find that more boys are affected (19–30% of referrals) than is the case for adult males (5–10% of cases).

There is some indication that in adolescence, anxiety about eating is even more common than concern about weight. In a study of 462 adolescents, Mueller et al. (1995) found that while 10% reported themselves as 'underweight' and 21% as 'overweight', altogether 50% expressed anxiety about over- or undereating. There was no correspondence between the concern about eating and whether the adolescent's weight varied in that direction. The suggestion that adolescents are worrying inappropriately is supported by the fact that none of the sample were known by their teachers to have an eating disorder.

Advertising food to children therefore takes place against a background of both justified and exaggerated anxieties of both parents and the young people themselves.

1.6   The processes of food choice

Food choices emerge from processes that take place over time and are affected by a variety of factors. A choice may occur at the time of selecting foods for a meal; while shopping; while planning shopping; in response to a request from a family member; in response to events during a meal, such as a refusal to eat a food category; while being exposed to a new food (in an advertisement; when eating out, or on holiday); and many other situations.

Each point of choice will also be affected by a variety of considerations. One set have to do with getting the children fed, but these must fit in with other objectives. Our research has shown how varied these objectives may be, including: minimizing cost; fitting into routines; having a meal that provides a time for conversation; bribes and rewards; proving what a good parent you are; using meals to teach manners; minimizing effort; an assumption that the parents will eat whatever the children have chosen; a rule that cooking that makes pans dirty is only done at weekends; and sometimes even enjoyment of food.

When considering the possibilities of influencing food choice it is therefore crucial to know at which point in the decision-making process the influence will operate. Then it is necessary to have an understanding of the other factors operative at this stage and how the influence will be carried forward to the purchasing occasion. It is also relevant that much food is chosen and consumed without the need for substantial decisions. The time of day and type of meal lead to certain assumptions about what is appropriate. Birch, Billman and Richards (1984) found that children as young as 3 years had clear expectations about what was appropriate for breakfast and dinner. Also, many decisions are made at the time of shopping and then leave no requirement for choice at the time of a meal. If breakfast is cereal and there is only one kind of cereal in the cupboard, there is no need for much thought about what to have for breakfast. Influences on food choices are therefore likely to be quite dissociated from the point at which the choice is made. Television advertising, unless it triggers a search leading to immediate consumption, will be separate in time and place from both purchasing and consumption decisions. In-store advertising and packaging operate directly on purchasing decisions but these may be highly constrained by earlier events in the home, such as preferences and refusals expressed by the children.

Influencing food choice may therefore encounter difficulties under three headings:

- the complex (sometimes competing) agendas that the provision of food in the family must satisfy;
- the fact that the point of maximal influence may be a long way away from the point of decision;
- the fact that much of the behaviour is not easily open to any kind of influence.

This preliminary survey shows that any attempt to influence children's diets, for example by advertising, involves attempting to add one influence into a complex process. What a family chooses to purchase, and what is chosen for eating on a specific occasion, will be affected by a whole range of factors. Informed judgement alone will not be an adequate guide in such a complex situation. Only through detailed research will it be possible to chart the factors relevant to a particular product and map out the potential consequences of different strategies. However, the degree of complexity indicated by the overview of factors in food choice offered above, poses a challenge for research. It is extremely unlikely that an adult or a child would have enough insight into the processes by which they choose food to be able to describ them to an interviewer or rate them on a questionnaire. In such circumstances it is necessary to use a technique which will obtain data from family members about the full range of experiences they have had regarding making food choices, and to use these data to tabulate the influences involved. An example of such research, which was designed to discover the pattern of belief within families, follows.

1.7   Recent research into family food choice

Extensive research into food choice has been carried out by The Psychology Business both as strategic and as commercial projects. The same basic methodology has been used in all of the studies, and is described briefly here. More detailed accounts of the methodology are available (Stratton, 1995, 1996) and a full report of one of the major studies 'Influences on Children's Diets' is available from the Advertising Association (1994).

The core of our approach is to gather very detailed information about the concerns of family members. This necessitates a qualitative approach with a system of interviewing which has been specifically designed for use with families. The systemic approach to family therapy offers powerful methods for helping families to articulate their beliefs and concerns. We have, therefore, used a well-documented form of this approach (Stratton, Preston-Shoot and Hanks, 1990) and adapted it to provide a standard system for interviewing individual family members and whole families.

In the research commissioned by the Advertising Association (AA), for example, we interviewed all members of the family individually (and privately) and then the whole family together. When there were children less than 6 years old the whole family was only interviewed together. We also included eight groups of mothers. This mixed procedure provided valuable information about the positions that would be taken by individuals when removed from the influence of their family. Other studies have ranged from large samples interviewed individually through to multi-method approaches to small samples of particular populations with matched controls. In the latter type of research it is possible to use intensive techniques such as filming families at mealtimes to supplement the other material.

When dealing with a large volume of qualitative material it is essential to have a coherent system to process the information. We have chosen to base our system in attribution theory (Hewstone, 1989) because this is the most highly developed approach to analysing the belief systems by which people guide their actions. The method depends on identifying all explanations and statements about cause and coding these under multiple headings. Attributional research has shown that people are most likely to offer explanations (attributions) around issues that concern them. We can, therefore, use the relative frequencies of attributions as an indication of the significance of different issues. We can also directly compute the numbers of times different causes are offered to account for given outcomes. Let us consider some examples of what this approach generated in the AA research.

1.7.1 Family dietary priorities

In all, over 7000 attributions were identified, recorded in verbatim form and coded. The first general question was to identify the broad issues discussed by families, and to discover their relative importance. Attributions containing any reference to food were selected and classified under four broad headings as shown in Fig. 1.1.

It can be seen that practical issues to do with the provision of food, and issues of food choice are the two dominant priorities. Whether or not food is enjoyed is important but receives less than half of the attention of the two major factors. Explicit references to nutrition and health are even less frequent. The 7% represents a total of nearly 500 statements, and there is no doubt that nutritional concerns were implicit in some of the other accounts. Even so, the data give a clear indication that nutritional concerns are not frequently at the forefront of family food choices.

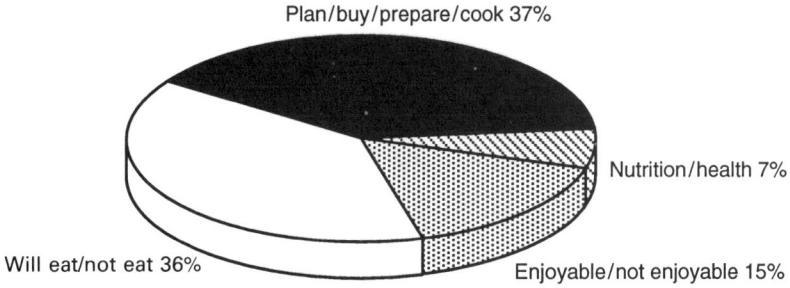

Figure 1.1 Proportions of attributions on broad definitions of family dietary priorities ($n = 6826$).

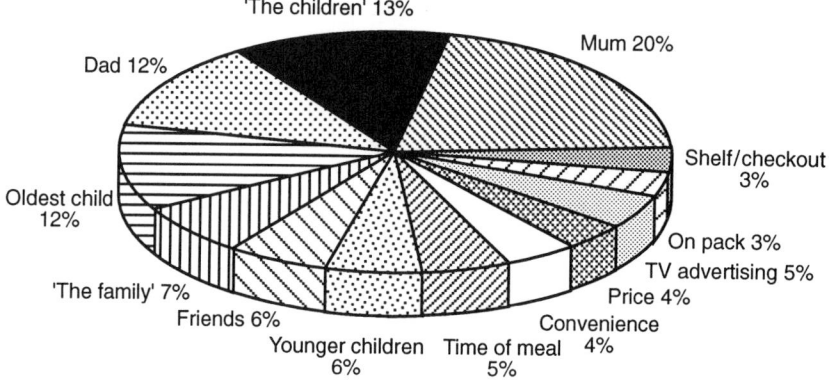

Figure 1.2 Proportionate emphasis by families on influences on their food choices ($n = 3004$).

1.7.2   Influences on family food choices

A central issue is the frequency with which different sources of influence are used by the family to account for their food choices. We therefore selected those statements in which some kind of food choice was the outcome ($n = 3004$) and classified the causes that were offered (Fig. 1.2).

This data set launched us towards our major conclusions. Although much of the data pointed to the dominant role of the mother, this set showed that in 31% of those statements directly concerned with food choice, one or more of the children were seen as responsible. External influences make up at most 30% of the total. Families therefore see an approximately equal distribution of direct influences on food choice between the parents, the children and outside influences.

1.7.3   Priorities for parents and children

The next issue is of whether the major priorities differed between parents and children. Accounts were grouped into four major headings and their frequencies are shown for parents and children separately in Fig. 1.3.

As might be expected, there is a somewhat greater tendency for practicalities to be applied to the activities of parents, while the issue of whether the food is or is not eaten occurs more frequently in relation to the children. Even nutrition and enjoyment apply more often to parents than when the children are the main focus of discussion. However, the overall impression from Fig. 1.3 is the similarity of concerns irrespective of whether it is the parents or the children being discussed. Because Fig. 1.3 was derived from the full data set, the majority of attributions will have been provided by adults, so we are primarily reporting the adult perspective. When comparisons were made between child and parent

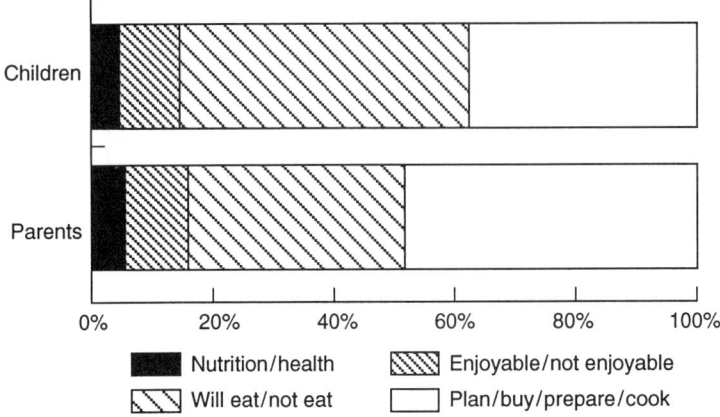

Figure 1.3 Family dietary priorities for parents and children (n = 2048 Parents; 2106 Children).

attributions it was again the similarity rather than any major differences that was striking.

1.7.4   The role of different food types

When we unpack these categories in terms of the types of food involved, we do not see the pattern that would be predicted by groups who are greatly concerned about the effects of advertising in directing children towards sweet or fatty snacks. Fig. 1.4 plots the main family issues in relation to type of food.

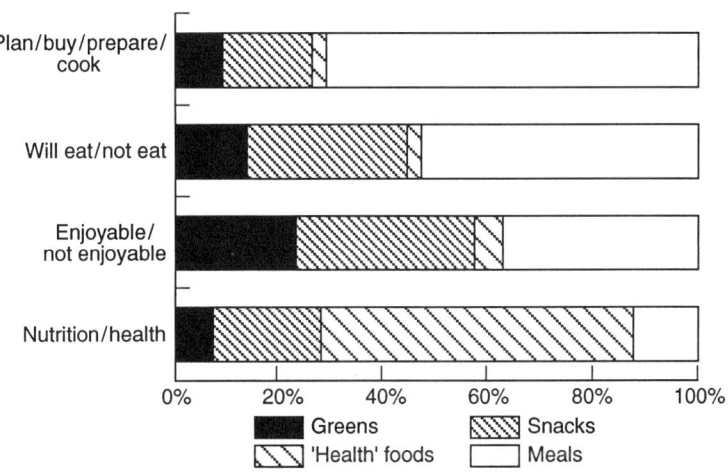

Figure 1.4 Role of different food types in major food issues (n = 2672).

The four broad family priorities on diet scaled to 100% show that activities around planning, buying, preparing and cooking are talked about in terms of meals rather than individual food types or components. Fruit, vegetables and salads are always prioritized less as an issue than either snacks or meals, though they are talked about to some extent in terms of whether or not they are enjoyable, e.g. 'they are all soggy, I don't like the school cooked carrots'.

Not only does nutrition and health only account for 7% of family priorities, but the major threat to dietary balance is that families think about nutrition predominantly in terms of specialized 'health' foods and supplements rather than in terms of the everyday meals, snacks or fruit and vegetables they are eating.

Snacks often tend to be seen as outside of the nutritional process. As food they are used to keep children going between meals, then they are used as something nice to have in regular situations such as coming home from school, and also as treats. All of the tabulated data presented here come from the AA study but other research using similar methods has found that an appropriate and regular use of snacks correlates with positive attitudes to diet and to the enjoyment of food.

For research purposes we have taken the family's own definition of a snack. For marketing purposes it is important to recognize that these definitions may not correspond precisely with industry assumptions. For some families a ham sandwich was regarded as a snack. For others, only sweets and crisps counted.

1.7.5   Differentiating parent and child behaviours

For the next analysis we took those incidents of food choice which were caused by a specific behaviour of either a parent or a child. Parental behaviours were twice as common as child behaviours, but the distribution, as shown in Fig. 1.5. was different.

In Fig. 1.5 we see a progression in which parental influence is very strong around practicalities, but the influence of the child becomes progressively more important as issues of choice come into play. In absolute terms the major issues are the household routines, developmental aspects such as nutritional requirements and changing tastes. None of these will be easily changed by outside influences. The next rank includes resistances, the issue of whether the child will eat all of what is offered, what they say they would like, and also television. This last factor is split between influences taken from television and, rather more strongly, the effect of watching TV on the patterns of eating. We could label this second rank of influences 'elective factors' because they are, with the exception of developmental factors, around areas in which the child has a certain amount of freedom of choice. Being elective, they are also open to influence.

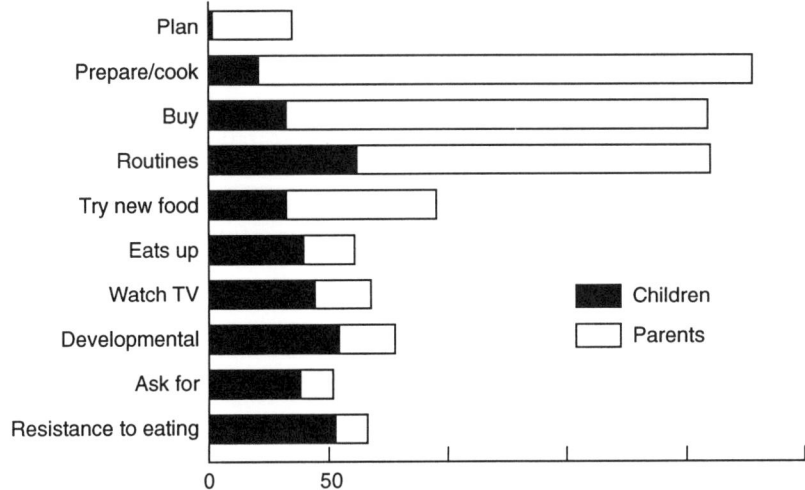

Figure 1.5 Parent and child behaviours influencing family food choice ($n = 725$ for Parents; 376 for Children).

Figure 1.5 encapsulates the major tendencies in family food choices. The strongest issues are the practicalities of budget, family routine and what is involved in preparation etc. After these parentally-driven issues are a set around the basic issue of giving the children what they will eat. The data analysis showed clearly that the most significant influence from children was the parents' judgement about what they would eat. Food planning was around the most practical solution to obtaining and preparing food that would be eaten. This was not presented by parents as a contentious issue in which they felt 'pestered' to give their children inappropriate foods. It was much more about maintaining enough variety in the face of their children's likes and dislikes to stop them becoming bored. It must be relevant in this process that very many parents reported that their own diets were strongly influenced by what they provided for their children. Even if they did not eat with the children, they were likely to prepare one basic meal that would be eaten with variations by all of the family. Variety and interest for the children therefore meant variety and interest for the parents as well. This could be one reason for the success of new foods such as pizzas which are as acceptable to adults as they are for children.

1.8 Implications of the research

In pulling together the substantial body of data provided by the AA study and relating it to other research we have conducted, we found that our findings called into question a whole range of assumptions. These assumptions are not only at the basis of the claims of pressure groups such as the National Food Alliance (Dibb, 1993), but also seem to be widely accepted within the food and marketing industries. We have chosen to express these beliefs in a rather strong form as 'myths' about attitudes to food, food choice and advertising, within families. Each of the six myths is challenged by our research findings, and the six alternative perspectives suggested by our research together make up a useful basis from which to consider the role of the food industry.

1.8.1   Myth 1

Parents see themselves as gatekeepers, battling to keep out foods they consider unhealthy for their children

The parents we interviewed seem to have a more rational attitude which is that no food is intrinsically unhealthy. They do not see their job as keeping any particular foods out, but as ensuring that as wide a range as possible is offered to their children.

1.8.2   Myth 2

Families see advertising as distorting the pattern of their children's eating

Our parents did not reveal a generally negative attitude to advertising. They were well aware that children's food advertising concentrates on snacks, drinks and cereals but did not see these categories as distortions of their children's diets.

1.8.3   Myth 3

Parents feel powerless to control their children's nutrition

Apart from creating a disparaging view of parents, this belief suggests that whatever children decide to eat will be bought for them. This would be seriously misleading both to the industry and to outside bodies. Both children and parents see food choice as something that is negotiated within the family. While some families feel highly constrained either by the range of what they are prepared to consider or by finance, in most families food choice is an active interactional process.

1.8.4   Myth 4

Parents believe television is the major influence on their children's diet

Parents see television as a relatively minor influence on what the family eats. Since they are not asking for their children to be protected from television, we would be better finding out from them what they **do** think influences their children's diets, and respecting parents' understanding of their own children. Improving the 'Health of the Nation' requires a concerted effort to improve the Nation's diet. This effort could be seriously misdirected if resources are diverted in unhelpful directions.

1.8.5   Myth 5

Parents see children as continually pestering them for specific advertised foods

A major tendency in the families interviewed was the positive tone in which food preferences were discussed. At the worst, parents sounded resigned to the peculiarities of children's preferences, but they showed no signs of resentment about children expressing preferences. This myth would feed into a view that children should be prevented from having influence over their diets – a view that we believe to be profoundly mistaken. It would also give a very unrealistic idea of the interactions around food choice which are an important part of the life of ordinary families.

1.8.6   Myth 6

Parents worry that their children are only interested in advertised junk food

There is an underlying assumption in this myth that what is junk for adults is junk for children, which has been discussed previously. Parents do not generally claim that it is important to reduce children's interest in junk food. They see fast food and snacks as helpful in the overall pattern of eating.

1.9   How advertising can fit into the food choice process

This chapter has attempted to provide a soundly grounded account of the complex issues involved in family food choice. From this account it is now possible to review the role that advertising might play, and the factors which will determine whether advertising is effective.

We should return to the issue of the stages of the food choice process. A starting point is any occasion on which a family member encounters information about a food they might like to try for the first time or return to after a gap or increase/decrease their consumption of. Children are often the bearers of such information, though fathers are particularly likely to make suggestions for more exotic (foreign/highly spiced/expensive) items. Information about new items often comes from television but equally from friends, or when eating outside the home. While mothers dominate food choice decisions they are perfectly willing to be influenced by their children, and children do request foods that they have seen advertised (Taras *et al.*, 1989). How much this affects food purchase and consumption must be determined by the relationship between the food requested, the place of the request in the sequence of decision-making processes, and the potential role of the requested food in the family pattern of eating.

Our data show that families regard advertising as one among many influences, and do not, in general, have a negative reaction to this influence. Certainly, advertising influences children who then request foods they have seen in the advertisements (Taras *et al.*, 1989). Donkin, Neale and Tilston (1993) found that 7 to 11 year olds were especially likely to request advertised foods that they could eat on their own. The confusion comes if it is assumed that parents will view such requests negatively. There seems to be no basis for this assumption, and therefore no basis for automatically equating a child's request for, or interest in, a particular food as 'pestering' (Stratton, 1994).

There are concerns about food advertising relating to the form and context of the advertisements. Rajecki *et al.* (1994) analysed 92 (American) food advertisements aimed at children and found the most common feature to be violence, followed by conflict. With older children it is the potential association of advertising content with eating disorders which worries observers most. Ogletree *et al.* (1990) found that while most food adverts featured male figures, when the emphasis was on appearance, there were significantly more female characters. Myers and Biocca (1992) report a study in which a relatively small exposure to videos promoting the slim ideal body were enough to change the body image of young women. Magazines probably promote thinness even more strongly than television advertisements. Guillen and Barr (1994) tracked a magazine for adolescent girls over 20 years and found an increasing majority of material in which nutritional issues were presented in terms of thinness, while Hertzler and Grun (1990) examined 117 magazines and found an implication that women need to be slim, as well as fit and young, and to use cosmetic products in order to be beautiful.

Food advertising is always going to operate with the primary objective of increasing market share. However, in this sensitive area it certainly makes sense to consider wider implications. The evidence reviewed in this chapter suggests that many concerns are overstated and some are entirely unrealistic. This does not mean that they can be disregarded. It is also suggested that the available evidence indicates that the greatest potential for damaging effects comes from incidental aspects of advertising. Aggression in boys and thinness in girls are the obvious examples. These are not intrinsic to most of the foods being promoted. There is also some indication that advertising encourages solitary eating among children.

Given the richness of psychobiological functioning, social learning and family interactions to which eating contributes, there are many other aspects to which advertising could be directed. Some of the benefits of food advertising may only be realized if there is a shift of emphasis from brand promotion to product promotion. It is tentatively suggested that criticism of food advertising to children has produced a defensive atttitude in which it has seemed important to claim that the advertising is directed at the market share of the advertised brand without changing product choice. In fact, one of the most positive contributions that could be made by the industry, and one that would be welcomed by most parents, would be to extend the range of products that children will consider.

1.10 Conclusion

Advertising influences food choice. The evidence reviewed in this chapter suggests that this is not, in itself, a cause for concern. The interacting influences on children's eating are rich and complex, and advertising makes a small but significant contribution that is not generally regarded negatively by parents. However, the central role that eating plays in so many social and family processes means that any external influence needs to be handled with care.

Food promotion has the potential to have (or at least be seen to have) damaging consequences of two kinds. The first relates to nutrition. If food advertising is seen to increase the tendency for children to restrict their diets, and at the same time foster a reliance on foods with limited nutritional value, then the argument can be mounted that its overall effects are negative. The second concern is not with the foods advertised but the incidental messages that accompany the advertisement. The promotion of thinness for young women may well be a contributory factor to the increasing levels of elective undernutrition and eating disorders in this group.

The industry will continue to use those images which appeal to children. But it must be concerned not to do so in ways that feedback to give undue strength to images that are dangerous. Insofar as children's food advertising can be characterized as promoting aggression among boys and unrealistic thinness among girls, it is open to criticism. Hopefully this review has indicated that the range of potential influences on food choice within the family is much wider and offers plenty of scope for imaginative products, packaging and advertising which increases rather than decreases the contribution of eating to the richness of children's lives.

The myths described in this chapter arise from a negative and restrictive set of attitudes. An alternative perspective is that television viewing, of which advertisements are an important part, is now a major component of the cultural and social lives of children. There is no issue of whether this can be reversed, so it must be used positively. Food makes a major contribution to the richness of child and family life, and often confers a positive tone and value to other important activities. If food advertising can incorporate and reflect this richness and positiveness it will be playing an appropriate role in the lives of children.

References

Advertising Association (1994), *Influences of Family Food Choice.* Research by The Psychology Business. AA, London.

Birch, L.L. (1980), The relationship between children's food preferences and those of their parents. *J. Nutritional Education*, **12**, 14–18.

Birch, L.L. (1987), The acquisition of food acceptance patterns in children. In *Eating Habits, Food, Physiology and Learned Behaviour*, Boakes, R.A., Popplewell, D.A., and Burton, M.J. (Eds), John Libbey, London, 107–130.

Birch, L.L., Billman, J. and Richards, S. (1984), Time of day influences food acceptability. *Appetite*, **5**, 109–112.

Blundell, J.E. and Bauer, B. (1994), Eating disorders in relation to obesity: semantics or facts? In *Obesity in Europe*, Ditschuneit, H., Gries, F.A., Hauner, H., Schusdziarra, V. and Wechsler, J.G. (Eds), John Libbey, London.

Bryant-Waugh, R. and Lask, B. (1995), Annotation: eating disorders in children. *J. Child Psychology & Psychiatry*, **36**, 191–202.

Dibb, S.E. (1993), *Children: Advertisers' Dream, Nutritional Nightmare?* National Food Alliance, London.

Donkin, A.J.M., Neale, R.J. and Tilston, C. (1993), Children's food purchase requests. 1st Food Choice Conference (1992, Brussels, Belgium). *Appetite*, **12**, 291–294.

Erikson, E.H. (1963), *Childhood and Society.* Norton, New York.

Guillen, E.O. and Barr, S.I. (1994), Nutrition, dieting, and fitness messages in a magazine for adolescent women, 1970–1990. *J. Adolescent Health*, **15**, 464–472.

Hanks, H. and Hobbs, C. (1993), Failure to thrive – a model for treatment. *Bailliere's Clinical Paediatrics*, **1**, 101–119.

Harper, L.V. and Sanders, K.M. (1975), The effects of adults' eating on young children's acceptance of unfamiliar foods. *J. Exp. Child Psychology*, **20**, 206–214.

Hertzler, A.A. and Grun, I. (1990), Potential nutrition messages in magazines read by college students. *Adolescence*, **25**, 717–724.

Hewstone, M. (1989), *Causal Attribution.* Blackwell, Oxford.

Martin, M.C. and Kennedy, P.F. (1993), Advertising and social comparison; consequences for female preadolescents and adolescents. Special Issue: The Pursuit of Beauty. *Psychology and Marketing*, **10**, 513–530.

Mueller, C., Field, T., Yando, R. *et al.* (1995), Under-eating and over-eating concerns among adolescents. *J. Child Psychology & Psychiatry*, **36**, 1019–1025.

Myers, P.N. Jr. and Biocca, F.A. (1992), The elastic body image; The effect of television advertising and programming on body image distortions in young women. *J. Communications*, **42**, 108–133.

Ogletree, S.M., Williams, S.W., Raffeld, P. *et al.* (1990), Female attractiveness and eating disorders: Do children's television commercials play a role? *Sex-Roles*, **22**, 791–797.

Poskitt, E.M.E. (1994), The prevention of childhood obesity. In *Obesity in Europe*, Ditschuneit, H., Gries, F. A., Hauner, H., Schusdziarra, V. and Wechsler, J.G. (Eds), John Libbey, London, pp. 141–145.

Rajecki, D.W., McTavish, D.G., Rasmussen, J.L. *et al.* (1994), Violence, conflict, trickery, and other stories in T.V. ads for food for children. *J. Appl. Social Psychology*, **24**, 1685–1700.

Stratton, P. (Ed.) (1982), *Psychobiology of the Human Newborn.* John Wiley & Sons, Chichester.

Stratton, P. (1994), The myths about children's dietary choices. *Admap*, December 94, 20–24.

Stratton, P. (1995), Systemic interviewing and attributional analysis applied to international broadcasting. In *Psychological Research: Innovative Methods and Strategies*, Haworth, J. (Ed.), Routledge, London. pp. 1–14.

Stratton, P. (1996), Attributional coding of interview data. In *An Introduction to Qualitative Methods* Hayes, N. (Ed.), Routledge, London (in press).

Stratton, P., Preston-Shoot, M. and Hanks, H. (1990), *Family Therapy: Training and Practice*. Venture Press, Birmingham.

Tanner, J.M. (1978), *Foetus Into Man*. Open Books, London.

Taras, H.L., Sallis, J.F., Patterson, T.L. *et al.* (1989), Television's influence on children's diet and physical activity. *J. Developmental & Behavioral Paediatrics*, **10**, 176–180.

Valsiner, J. (1987), *Culture and the Development of Children's Action*. John Wiley & Sons, Chichester.

2 How to make effective advertising aimed at children

J. MATHEWS

2.1 Introduction

Given the number of new food and drink products introduced each week –
one report cites over 80 in the UK alone – it is understandable that many
manufacturers feel daunted by the task set them. How do they make their
products stand out from the crowd – especially among that most
demanding and volatile audience of children? Indeed, accepted wisdom is
that the child audience (in this context 6–12 years old) is wildly
unpredictable. What's 'in' one minute is 'out' the next. Consequently, it is
assumed that communicating with them via advertising, packaging,
promotions, p.r., or indeed, the development of new food and drink
products, is a highly risky business. I simply don't believe that to be true.
On the contrary, I believe there are certain triggers or basic motivations
that will always strike a chord with children. When these 'enduring themes'
are used, and delivered in an appropriate way, provided the product has a
role to play in the child's life, the commercial (or promotion etc.) will
almost certainly work. It is certainly possible to be successful with children,
not just by chance, but by design. In addition, these triggers have proven
themselves (in J. Walter Thompson's experience) to be relevant to children
around the world. Their basic motivations are the same everywhere.
Although I have written this chapter specifically with food and drink
manufacturers in mind, the principles hold true for all advertising to
children, whatever the category.

By the end of this chapter, you should be in a better position to judge
whether a new commercial (or product, promotion or character) is likely to
be successful with children and, more importantly, be able to articulate
why this is so, and use this as a guideline for future development.

2.2 Understanding the audience better

'Thou shalt understand thy target audience'

is the standard phrase trotted out to brand managers and advertising account executives wet behind the ears. And, indeed we feel that we do have an understanding of children – after all, we were all young once. However, our experiences as children don't equip us for an understanding of children today. When you learn that the average American child has watched 8000 televized murders before finishing elementary school, or that the majority of children have a TV in their bedroom, and they can all use computers with the aplomb of an expert, you begin to realize that a conscious effort has to be made to learn about these children.

If you can't name the top selling toys, the favourite TV programmes, music, clothes, language or pastimes, or remember when you last sat in on some qualitative child research, you are out of touch. How do you go about finding out more?

My advice is to read everything you can about children in books like this, and every article you can lay your hands on, go to toy fairs etc. But much more importantly, go out and find out for yourself. Before developing their very successful 'Wild Creatures' campaign in the USA, Levi's poked around in children's wardrobes, and followed them on shopping trips to really understand their audience better. In a similar vein, Fisher Price run a school in the USA where some of their toy concepts are tested, and Sega in the UK even have a junior 'board' of children who are consulted on new product development and advertising issues. These companies listen.

At J. Walter Thompson in London, we've been running a 'Kid Panel' consisting of the same 60 children, who we interview once a year. While undeniably not a representative sample, it has helped us to explore how trends evolve among the same group of children, and specifically explore why some themes last longer than others. This helps us to **predict** rather than react. Most information about 'What's in' for children is *per se* out of date by the time its received, and positively fossilized by the time it can be used for marketing to children.

Why introduce a dinosaur-shaped food product a year after the craze has passed? Try to tap into the basic motivations (outlined later in this chapter) rather than passing fads. I am constantly amazed, given how relatively easy and inexpensive it is to set up a context in which to talk to children regularly, how few manufacturers do it. Think about setting up your own 'kid panel' to ensure constant dialogue – there's nothing more sobering than having your beloved concept dismissed out of hand by a group of 7-year-olds as being 'boring' or 'too young' for them. We've often asked children to help us design new products such as a new breakfast cereal, or ice cream. The results have often been a revelation. The ideas themselves are rarely practicable – but look behind them: an unusual combination of colours and textures, or a desire for a miniature 'kid size' product. This is when listening to children really pays off, and gives you an edge over the competition.

2.3 Ages and stages

I cannot overstate how important it is to understand how children change at different ages. A thirty- and forty-year-old adult have more in common than a six- and eight-year-old child. An advertising audience that refers to '6–15 year olds' should be outlawed! Incidentally, we have noticed there is a particularly marked watershed when children change schools, and there is still a huge gulf between boys and girls.

4 years	$\longrightarrow$	12 years
Magical	$\longrightarrow$	Rational
Physical	$\longrightarrow$	Emotional

Broadly, you can say that children move from the magical, physical years to be more rational as they mature. To some extent, a child's reaction to advertising also becomes more rational as they mature, although their primary reaction is still **emotional**.

We've divided the children into 3 groups, looked at some of their key traits, and how their response to advertising differs by age:

- 'Explorers': 4–6 years;
- 'Conquerors': 7–9 years;
- 'Groupies': 10–12 years.

Figures 2.1 and 2.2 summarize our findings.

Explorers (4 – 6 years)
- There's so much to discover.
- Self-centred, egocentric.
- Parental authority unquestioned.
- Able to identify packs.
- Love fantasy.
- Boys – boisterous action.
  Girls – cute and pink.
- Very literal.

Conquerors (7 – 9 years)
- Plenty of confidence.
- Logic begins to develop.
- Passion for possessions.
- Product requests reach peak.
- Girls vs. boys.
- Eager to grow up.
- The peak of childhood.

Groupies (10 – 12 years)
- Peer group rule OK.
- Very sensitive to how others see them.
- Develop independent judgement.
- Look for rational reasons why.
- Boys – mechanical wizards, computers, sport.
- Girls – fashion and giggles.

Figure 2.1 A segmented audience. (© J. Walter Thompson).

Explorers (4 – 6 years)

- Consider advertising as entertainment to be enjoyed vs. understood.
- Can distinguish between advert and programme.
- Short attention span.
- Enjoy repetition.
- Notice small details, and repeat them afterwards.
- Love animation.

Conquerors (7 – 9 years)

- Absorbed in TV.
- Understand that adverts are to persuade people to buy.
- Discriminate between good and bad 'adverts'.
- Very high involvement with 'adverts' that are liked.
- Limited tolerance of advertising that shows people behaving unrealistically.
- As product requests reach a peak, advertising provides rationale for parental requests.
- Can follow storylines and plots.
- Enjoy visual humour (slapstick, adults looking foolish etc).
- Prefer conventional story structure – beginning, middle, end.

Groupies (10 – 12 years)

- Watch more adult programmes, and as a result may be more sophisticated about plots, themes, video techniques.
- Aware of advertising's function to sell products, and want to know more about product.
- Accept advertising, but with critical judgement ('It can be a con')
- 12-year-olds consider many children's adverts to be for 'young kids'.

Figure 2.2 How the response to advertising differs by age group. (© J. Walter Thompson).

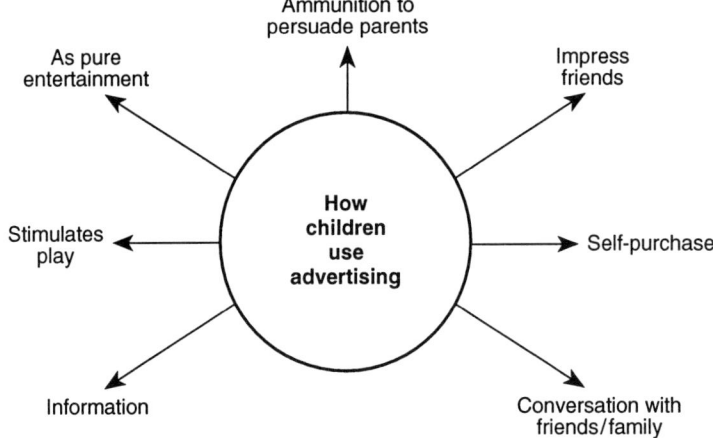

Figure 2.3 How children use advertising. (© J. Walter Thompson).

2.4   Advertising to children

In the UK, children see over 1000 TV adverts per month (half of the US total!). They love TV, and often cite it as their favourite activity. Advertising is also enormously enjoyed, and is used by the children in many ways (Fig. 2.3).

It is generally agreed that the influence children have on purchasing is becoming stronger. While their sphere of influence is growing to areas such as pet food (where they are deemed to be the experts) and brown and white goods, it is still food and drink – relatively low risk for the purchaser – where their influence is strongest. So how do we ensure that our advertising, which is such a critical part of the mix, is optimal?

The most successful advertising to children is a combination of:

(a) Enduring themes – universal motivations or emotional triggers guaranteed to strike a chord;
(b) Executional elements – presented to children in an attractive appropriate way.

2.5   Enduring themes

Children's primary response to advertising is emotional, and that is what enduring themes tap into. They are basic truths about what motivates children, and what they identify with.

We have analysed the most successful children's TV commercials, (and press adverts, TV programmes, films, toys etc.) and the same themes keep emerging:

1. kids in control;
2. wanting to be older;
3. social acceptance;
4. precious possessions;
5. Good vs. Bad.

2.5.1 Kids in control

Often children feel dispossessed and overwhelmed in the adult world. In advertising, you can turn that on its head, and children can be in control. What they want more than anything is their own money, freedom from parents to make their own choices, own room and so on. Reflect this!

There are different ways that 'kids being in control' can manifest itself:

(a) pulling the strings;
(b) feeling intellectually clever;
(c) mastering a skill;
(d) winning.

(a) Pulling the strings.   Children love to be the ones in control – especially if it involves getting one over adults. The success of 'Home Alone' is testament to this, as is the Pepsi 'Shaq' commercial, where the little boy refused to relinquish his drink to the basketball star. Kids – 1, Adults – 0!

(b) Feeling intellectually clever.   The point here is not learning but knowing. Prowess is more important than knowledge, as children desperately try to impress their friends and family with what they know. So offer children little nuggets of information in bite-size packages (the computer games magazines and some children's TV programmes do this very well). Make it easy for them to learn the information to allow them to show off to their friends . . . like baseball cards do. Give them some ammunition in the form of weird facts or jokes on your packaging.

I'm sure the desire to be seen to be intellectually clever helped fuel the dinosaur craze as a child could turn to the mother and say with exasperation 'It's not a brontosaurus, it's a brachiasaurus, Mum'. In the same way, know the rituals surrounding the Power Rangers transformations or the X Men's powers.

(c) Mastering a skill.   Children love seeing people – especially other children – mastering a skill, and they love a challenge. At a young age its mastering a construction set or a cartwheel, making their own sandwiches, using a magic toy set, etc., moving to sport and computers as they get older

(Kraft ran a successful 'design a sandwich' competition for kids in the US). I think sport, in particular, offers huge opportunities for advertisers. Being fit is definitely cool. Sport, incidentally, along with music and films, is a key international theme. McDonalds in the US tapped into this theme extremely successfully, using Michael Jordan in their advertising – someone whose skills (displayed in a suitably over-the-top fashion in the advertising) are desperately admired by children.

(d) Winning.   The ultimate proof of being in control is winning. What makes this particularly relevant for children is overcoming an obstacle or challenge to win. You've got to earn it.

   Kellogg's 'Frosties' advertising with Tony the Tiger is a great example of this and is used successfully the world over. (Video games also fulfil this, as each go you overcome more challenges to reach your goal.) You can either win by strength and practice or you can show ingenuity, which is also powerful with children. (In the film 'Benny and Joon' children loved the ingenious way they made a toasted cheese sandwich with an iron.) One Dairylea advert ('Kids will do anything for Dairylea') showed a child walking across the ceiling 'Spider-Man' style to reach the cheese. The children loved the ingenuity in that, especially because an adult had been outwitted en route!

   One final word on being in control. Its always best to have an audience to validate the mastery.

2.5.2   Wanting to be older

This is one of the classic enduring themes. All kids want to be older. This manifests itself by:

(a) doing adult things;
(b) imagining you're older or someone else.

(a) Doing adult things.   From an early age children copy their parents or siblings – which is why portable phones are the latest 'must have' accessory in the US, with models available for 18-month-old children upwards. Bikes offer children the chance to be independent. For them, they're like a magic carpet – their version of a car. They imagine they're doing adult things or being an adult. Don't give kids 'kiddy versions' of adult products. Make them as close to the adult product as possible – at least superficially.

   If they can avoid it, children will never play with younger ones, so the old adage about 'if in doubt, cast older' is absolutely right (along with the one about 'if in doubt cast a boy', as girls will relate to boys in advertising, but never vice versa).

(b) Imagining you're older or someone else. Kids like aspiring to characters they see on screen. 'I want to be like that' is a common feeling, and one that Fruitella tapped into brilliantly with their spoof of 'I'm Too Sexy' by Right Said Fred. The hero boy was seen singing 'I'm too juicy' in several humorous scenarios – often pitted against adults. Children loved it, and this success seems to have been reflected in the market.

Thus, you should do everything you can to avoid advertising that's 'for kids':

- children love being the smart ones – in sitcoms, children are adults and adults are children, e.g. 'Roseanne', '2.4 children'!
- don't show kids as kids – show them as being smarter, in control, clever.

Another way of exhibiting the desire to be older is imagining you're someone else. Hundreds of children must have imagined themselves having their own money, freedom from parents, being a rock star, or best of all, having their own place to do what they want.

This would, in part, account for the success of the best selling toy in history, with two sold every second of the day somewhere in the world. Yes, she uses roller blades, she drives a Ferrari, she saves endangered species, wears designer clothes, and wears big gold dangly earrings even when scuba diving. Yet it's BARBIE! Like many of the best toys, she's a prop for children's imagination – in this case, the child projecting herself using an aspirational character can be very compelling.

2.5.3 Social acceptance

This embraces three areas:

(a) having friends or being part of a group;
(b) impressing friends;
(c) not sticking out.

(a) Friends or part of a group. There's something very secure about being part of a group, and groups have appeared in children's lives from *Snow White and the Seven Dwarves*, through to the Beatles, the Brady Bunch, the Ninja Turtles, Thunderbirds, football teams, Take That, Power Rangers, X Men, Bike Mice from Mars (or as one child called them 'Cheese Chewers on Choppers'!) and so on. The list is endless. Children like the twin strands of security of the group, while also choosing their favourite individual within it. The anti-smoking lobby in the US have cleverly tapped into this, using the phrase 'No one wants to date a smoker any more'.

(b) Impressing friends.   There's lots of scope for promotions in this area, to give children the ammunition to impress their friends. It's especially important being the first one to have, or the only one to have.

(c) Not sticking out.   Although paradoxically, children like to think of themselves as individuals, they would hate to stick out in any way. There's enormous pressure on them to conform to the latest styles of clothes, music etc. They are defined, and define themselves, by their product choices. Give them a chance to be individuals, within an accepted structure – familiarity with a twist.

A successful Australian confectionery product, made by Nestlé, Yumbats (chocolate-shaped wombats), tapped into this enduring theme by basing its products and advertising on an irreverent animated 'gang', each with their own personality and foibles. Children identified with a particular character, and liked the security of the group. Initial sales were 160% over budget.

2.5.4   Precious possessions

A child's world is very small. When we asked children in the Kid Panel what was the most important thing that had happened to them in the last 6 months, they said 'I had my hair cut' or 'I got a new hamster', 'I've lost some teeth', 'I moved class'. School and home is it. Seen in this context, it's understandable how important their possessions are to them. Four areas of possession can be defined:

(a) ownership, hoarding, collecting;
(b) privacy, personal property;
(c) mothering, grooming;
(d) kid size.

(a) Ownership, hoarding or collecting.   Children love owning, organizing and gazing at their personal possessions. Little shelves and boxes are desirable.

(b) Privacy, personal property.   Children love to safeguard their posses-sions by hiding them away and personalizing them. Their bedrooms are their sanctuary (that's why tree houses are bliss) and offer manufacturers lots of opportunities to decorate their rooms, e.g. door hangers, posters, stickers etc. Get some photographs of your targets' bedrooms – there is almost no better way of getting to know them. See the contradictions of Blur posters on the walls and teddy bears on the bed.

Children also love secrets, like secret diaries (now updated electronically by Casio with My Magic Diary), and getting their own mail, joining clubs etc. (or finding something secret or hidden like the successful 'Harry the lime' promotion run by Ribena).

They love personalized things with their name, on especially stationery, lunch boxes etc., which help express their personality.

(c) Mothering, grooming.   Children, especially girls, love having pets or dolls to care for. Thus, you find the success of Cabbage Patch, My Little Pony, Sylvanian Families etc.

In fact, the appeal of these toys changes as the girls grow.

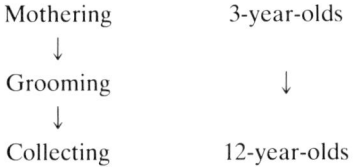

Mothering   3-year-olds
  ↓
Grooming   ↓
  ↓
Collecting   12-year-olds

(d) Kid size.   Children have a fascination with miniaturization – from *Gulliver's Travels* through to Puppy in my Pocket, Teddy Grahams, doll houses, juice boxes, Polly Pocket, Variety cereal boxes, Mighty Max etc. This would help account for the popularity of Kinder surprise, which is kid size. Think about the size of children's hands and mouths when designing products!

Little girls love the 'aaah!' factor in advertising and respond very enthusiastically to commercials like Andrex, Arthurs, Safeway and PG Tips.

2.5.5   Good vs. Bad

This is a classic and well-loved theme that has been around for years, from *Little Red Riding Hood* and *Cinderella* through *Superman* and *The Terminator* to the Power Rangers and Skeleton Warriors.

Children love to see the clash of good vs. evil, but would like the baddies to win now and again. They love being scared (which is different from being frightened). Spielberg got it right in *Jurassic Park* by using *T. Rex* and the *Veloceraptor* – the meanest dinosaurs – centre stage, and relieving the whole thing with humour too.

Boys love violence, and are preoccupied by destruction, super heroes, monsters, vehicles and weapons. The Teenage Mutant Ninja Turtles captured all of these – and tapped into wanting to be older.

This would account for the popularity of the vomiting Boglins which rejoice under the slogan 'They retch it, you fetch it'. Yes, children are fascinated by the human digestive system and the farting woopeeee cushion is still a strong seller. Figure 2.4 is a list of the top US candy products.

Don't be afraid to develop products which tap into this – children enjoy the tension inherent in this theme, and like to be challenged.

1. S.N.O.T. (Super Nauseating Obnoxious Treat)
2. Pop 'n Fizz Candy
3. Gummy Dinosaurs
4. Busted Gum
5. Bloody Sour Drops
6. Mad Dawg Super Spew Bubble Gum (You froth at the mouth when you eat it)
7. Spin Pops
8. Dinosaur Dust
9. Tongue Dippers
10. Tongue Splashers

Figure 2.4 US top ten candy products. (Source: NPD Inc.).

2.6   Executional elements

These enduring themes discussed in section 2.5 strike a chord with children. However, these alone are not enough. If the execution is not right, we may as well pack up and go home. A key thing to remember is:

Advertisements that fail with kids fail to entertain.

If they don't like your advert, if they're not attracted to it, it won't matter what you've said, because they won't be paying any attention.

We have identified three key executional elements:

1. interaction;
2. action/detail;
3. fun.

2.6.1   Interaction

This is the most important of the executional elements. By interaction I mean giving the children something to take away with them, something to copy. This is a way of extending your advert beyond 30 seconds and tapping into the jungle network of the school playground.

Interaction can take 3 forms – action, words and music (Fig. 2.5). Although, as a rule, *children remember what they see rather than what they hear*. Think of TV for children as being like a fireworks display. Actions speak louder than words. In the words of a 9-year-old boy 'You sorta listen with your eyes'.

(a) Action.   Children love to copy special actions, e.g. cool handshakes, funny walks. Tango tapped into this brilliantly with their orange genie slapping the Tango drinker (the brief was 'The drink with a hit'). As

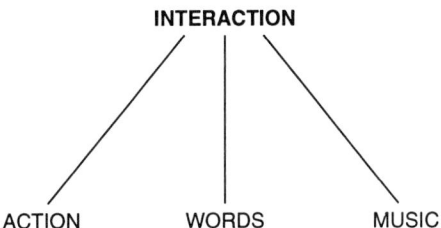

Figure 2.5 Interaction can take 3 forms.

testament to its success, the commercial had to be re-shot to replace the slap with a fake 'kiss', because children were copying the slap in playgrounds up and down the country, reportedly causing mayhem – not to mention a few burst eardrums. In the same way, kids have embraced the 'chicken flap' in the Chicken Tonight advertising, and it has proven to be (albeit unwittingly) a powerful reminder to the parents to buy the product. Like it or hate it, you can't ignore it!

(b) Words.   Younger children, particularly, are more likely to remember what they see in a commercial. As they get older they remember and respond to what they hear as well. Children love word play. Invent your own words, use repetition, jokes, mannerisms, a catch phrase or puns. Wotsits have run an excellent series of commercials that have tapped into this – 'Is he catching a bus or wot?', as a bus lands on a man. Very clever, and very successful.

Children's recollection of a voiceover is greatly enhanced if you do this, or use a child's voiceover, or a funny accent. And, please, avoid using 'kid speak'. It sticks out a mile. Any phrase that was 'in' when the script was written will sound horrendous when its aired. Playground credibility is granted, not claimed.

Make sure the diction of voices is clear and if you have a particular point you really want to get across, repeat it! Catch phrases at the end of commercials can be worth their weight in gold, and children often watch out especially for them. Popular examples include Milkybar, 'The Milkybars are on me!', Tango, 'You know when you've been Tangoed', Frosties, 'They're great' and Smarties, 'Only Smarties have the answer'.

Although this is a slight digression, while we're on the subject of words, I think manufacturers could be much more imaginative at making their products and flavour names more appealing. Take a leaf out of Monster Munch's book with their new 'Smokey Spider' snack variant. Children love it. Challenge yourselves with product and variant names – are they as intriguing, fun and catchy as they could be? Test them with kids!

(c) Music. Music is an integral part of the lifestyles of children, and a powerful advertising tool. Children will sing along, and this can't do anything but help the association.

Music has a broad age appeal, and 'in' music can bestow status on the brand – but beware of getting it wrong. You would do better developing something of your own – as Anchor butter have done so memorably with the singing cows, or Club biscuits have done with their 'If you like a lot of chocolate on your biscuit join our Club' song. Children up and down the country can repeat this verbatim.

Another excellent example of the use of music is in the enormously popular Maynards Wine Gums advert featuring a Scotsman dancing in a kilt, surrounded by dancing bagpipes, moose heads, Nessie in a bath and so on. The Scotsman says in a thick Scottish accent 'There's juice loose aboot this hoose!' Children love the unusual music, the details, the unusual accent. A great advert for them.

2.6.2 Lots of action/detail

Children have a very low boredom threshold and short attention span. If you've ever tried to entertain a pack of seven-year-olds, you would know that. The playground is a fast moving, exciting place. Reflect that. They are much more selective TV viewers than adults and are constantly channel surfing. Thus, if what they are watching doesn't grab their attention, they will switch. The same problem faces the marketers. How do you keep these children amused?

The answer is to keep moving! You will notice that children's programmes (like the Big Breakfast, Blue Peter, cartoons) and magazines are short, fast moving, with lots of different articles or scenes. Reflect this.

Lots of action includes keeping your product newsworthy – new promotions for McDonalds, a new persona for Barbie, a new shape of pasta or potato chip, a new colour in your Smarties. Scooby Doo discovered his nephew Scrappy Doo and Chip 'n' Dale became the Rescue Rangers. Successful marketers know that their message has got to stay fresh.

The recall level of children is three times higher than that for adults. Obviously this has implications on the way you time your media and on copy refreshment, i.e. shorter bursts, and frequent copy changes. If I were running a brand I would always ensure I had a pool of different executions running. There is nothing sacred about a 30-second time length. Think of developing more, different, shorter time lengths. In this case, its the sum of the parts that is greater than the whole. It makes the brand appear bigger than it is, and gives you a stature and presence in the child's eyes.

Children love small touches and notice lots of details – much more than adults do:

'. . . There are a few things you should do to make your cowboy programs (sic) more real. Have the six guns shoot only six bullets instead of about twenty before they reload. Have different trails. When a good guy chases a bad guy they go round the same rock about five times.

P.S. Let the bad men win once.'

(Letter from 7-year-old boy to USA Network.)

The best adverts should be rewarding enough to watch over and over again, like a good video game. The Smarties 'Inner Tube' advertisement is one of these. Children love the details – the 'fun' bits. Once we ran a 30-second animated commercial for Kellogg's Frosties, then, for the sake of media efficiency, cut it down to a 20-second commercial. We were careful to keep the plot and the product section intact, but cut out the 'fun bits'. When we researched it with children, they knew exactly what we had done and bitterly resented it. The answer is to keep the fun bits in, but ensure they are related to your product (e.g. the Chicken Tonight flap).

A useful device to start and/or end your advert with is to use something to catch the eye – a shout, or a visual trick. Kids have a way of cutting out the last 3 seconds of an advert (usually the boring pack shot) unless you give them a reason not to. Do something with the pack shot – incorporate it into the action, or animate it, or risk it being ignored.

Children do like to see what you are offering very clearly, and assume that if you don't show it, you have something to hide. However, given how important action and detail are, it's not always true that the more you put in the more you get out. Children can easily get overwhelmed by a surfeit of visual images. Ideally, they like a beginning, a middle and an end.

2.6.3 Fun

'It's funny' is the most common reason given by kids for liking an advertisement.

(*Source*: RBL)

Humour is one of the strongest weapons in an advertiser's armoury. Although children's sense of humour changes as they grow up (relying on verbal as well as visual cues) there are some things that will always work:

- slapstick;
- poking fun at authority figures;
- surprises;
- unusual walk or dress;
- funny costumes or disguises;
- incongruous things.

Some of the advertisements that children find funniest of all aren't aimed at them, e.g. the penguins in John Smith's bitter (incongruous, deadpan) or the Cook Electric animals (beautifully observed details). A recent advertisement for Pop Pop popcorn was also particularly well liked – the incongruous 'elephant bird', the clever end line 'the popcorn that goes ping!' and details like all of the animated children diving head first into the popcorn dish.

If you find a script funny, chances are, children will too. If you don't, chances are they won't either . . .

2.7 Deadly sins

Before concluding, it would be useful to summarize a few pitfalls. There are 'seven deadly sins' of advertising to children that must be avoided:

1. boring – lengthy dialogue, slow or complicated stories;
2. patronizing – underestimating children or pretending to be one of them;
3. out of date, especially in language;
4. unbelievable/stupid – stupidity would be adults dressing as children, for example. Thus, children will grant poetic licence required by any fantastic situation, but you have to play by the rules of that situation, e.g. Superman not using his powers;
5. relying on voiceover alone to deliver a message;
6. poor production values, poor casting (i.e. too young, too many girls, perfect children);
7. over-complicated (kids are more knowledgeable than we were, but not necessarily more mature).

2.8 Conclusions

In summary, it is my firm belief that every piece of communication we make for children can succeed. It needs a combination of understanding the target audience and what motivates them, and how to execute it (Fig. 2.6).

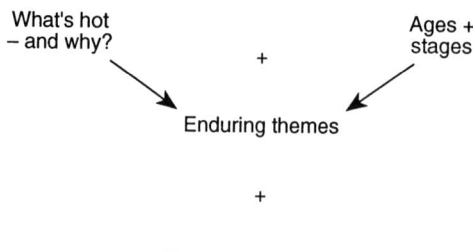

Figure 2.6 Understanding the audience.

- Challenge creative briefs to see if they show real insight into the target.
- Challenge the advertising to see if it taps into basic motivation.
- Challenge the execution to see that it has got true 'kid appeal' (and research it).

Although this chapter concentrates on advertising, manufacturers would do well to put their energy (via developing a deeper understanding of them) into developing products children really want, rather than finessing ways of selling to them. If your product doesn't suit and serve a child's life, start again.

I'm amazed at how few manufacturers really push themselves hard to understand this market better. Hopefully this chapter has given you a head start.

3 Starting the day right!
K.R. O'SULLIVAN

3.1 Introduction

Over the years people have made a lot of generous claims on breakfast's behalf. Experts and others have told us that it is not just a good way to start the day but that it is the most important meal we eat. They have told us that it makes us healthier, livelier, happier, fitter, brainier and, even, sexier.

But how much fact and how much fiction lies behind these assertions. Is a healthy breakfast really the best way for any person, but particularly for children, to prepare for the day or should it in fact be seen as nothing more than a minor snack?

This chapter sets out to put the scientific evidence under the microscope. Is breakfast really important for children, and if so what should they be eating? Can breakfast foods combat common nutritional problems of childhood and can breakfast benefit children not only in the short-term, but also have longer-term effects on their future health? Before these questions can be answered, we need to look at who does and does not eat breakfast.

3.2 Who eats breakfast?

It is no surprise to find that children tend to eat breakfast more frequently than teenagers and adults. Less than 5% of British (Kellogg's, 1992) and Japanese (Howden *et al.*, 1993) schoolchildren skip breakfast, compared to nearly four times more teenagers. In other countries, the incidence of skipping breakfast in all age groups appears very low, ranging from 1% in Indonesia, the Philippines and Singapore, to 13% in Thailand (Howden *et al.*, 1993). Interestingly, although nearly 85% of Koreans think that breakfast is important, 25–30% regularly skip it (Korean Dietetic Association, 1994). Some 29% of Canadians aged 15 years or older again regularly skip breakfast (Chao and Vanderkooy, 1989). While many who skip breakfast cite time and habits as reasons for this practice, if they realized just how important a meal it was, would they reconsider and start to eat breakfast again?

3.3 Break the fast!

As its name suggests, breakfast breaks the overnight fast which can last up to sixteen hours. It replenishes energy stores and gives a glucose boost (Pollitt, Leibel and Greenfield, 1981; Pollitt *et al.*, 1982/3). Refuelling with breakfast is important for children. The role of breakfast as an energy source is the one nutritional benefit most commonly recognized by consumers (unpublished report, J. Walter Thompson Company). Other benefits of breakfast have also been shown in studies carried out in many parts of the world looking at what happens when breakfast is skipped. Skipping breakfast has been shown to compromise the nutritional quality of the diet (Skinner *et al.*, 1985; Morgan, Zabik and Leveille, 1981; Morgan, Zabik and Stampley, 1986a, 1986b), have negative effects on certain health indicators (Stanton and Keast, 1989; Resnicow, 1991; Gibson and O'Sullivan, 1995), and has been shown to adversely affect mental and physical performance (Pollitt, Leibel and Greenfield, 1981; Meyers *et al.*, 1984; Simeon and Grantham-McGregor, 1989).

3.4 Compromising diet quality

Since breakfast supplies between 10–30% of the total daily energy and nutrient intakes it can make an important contribution to the nutritional adequacy of the total diet (Health & Welfare Canada, 1977; Martinez, 1982). Children who miss breakfast have lower intakes of important nutrients compared to their breakfast-eating peers. It appears that they often do not compensate for nutrients missed at this important meal. In a US study (Morgan, Zabik and Leveille, 1981) 5–12-year-olds who missed breakfast regularly, had lower intakes of iron, calcium, magnesium, zinc and vitamins A and B6. Often they do not meet the national nutrient intake recommendations for some vitamins and minerals (Nicklas *et al.*, 1993). Indeed breakfast has been found to be a significant source of these important nutrients.

These types of results suggest that breakfast omission in children may increase their risk of micronutrient deficiencies.

3.5 Breakfast and health status

Food choices made at breakfast may further impact not only the diet quality but also indicators of health status such as plasma cholesterol and body weight. Among High School children in the US, breakfast skippers have been found to have higher plasma cholesterol levels compared to those who ate some type of breakfast (Resnicow, 1991). Interestingly, in this study those eating eggs and bacon had lower cholesterol levels compared to breakfast skippers. However, children eating breakfast cereals had the lowest levels. Similarly in adults, those eating breakfasts that contain cereals tend to have lower cholesterol levels (Stanton and Keast, 1989). These results, together with those from other research (Jenkins *et al.*, 1989; Edelstein *et al.*, 1992; Arnold, 1993; McGrath and Gibney, 1994; O'Flaherty and Gibney, 1994) suggest that as long as people do not overeat, higher frequency of eating (for example, six small meals versus two large meals) may have beneficial effects on how the body metabolizes cholesterol.

In a study of breakfast eating habits in Hong Kong, schoolchildren skipping breakfast were heavier than those eating breakfast. In addition, skippers said that they felt tired and hungry significantly more often when they arrived at school, than those who had eaten breakfast (Guldon *et al.*, 1994). Gibson and O'Sullivan (1995) recently reported that British 10–15-year-olds who ate breakfast cereals had lower body mass indices (a measure of body fat) compared to their peers who did not eat cereal.

Given the public health significance of obesity and high plasma cholesterol levels as major risk factors for premature death from heart disease, developing and maintaining good breakfast habits in childhood should be an important objective in nutrition education (Gibson and O'Sullivan, 1995).

3.6   Mental, academic and physical performance

There is growing evidence that children may be more susceptible to the adverse effects of breakfast skipping on cognitive or mental function.

In a study carried out in the US (Pollitt, Leibel and Greenfield, 1981; Pollitt *et al.*, 1982/3) of 9–11-year-olds, breakfast consumption had beneficial effects, especially on children with low IQ, on some specific cognitive functions such as reaction time and problem-solving ability, when measured in the late morning. This work suggested potential overall improvement in school performance for children with low IQ. Similarly, in Jamaica (Simeon and Grantham-McGregor, 1987), the cognitive function of poorly nourished children was found to be even more vulnerable to missing breakfast than that of adequately nourished children. The omission of breakfast in Japanese students (Kagawa *et al.*, 1980) caused lower performance in aptitude tests, also absenteeism from class was lower in those who ate breakfast. These findings are illustrated in Fig. 3.1. Similar

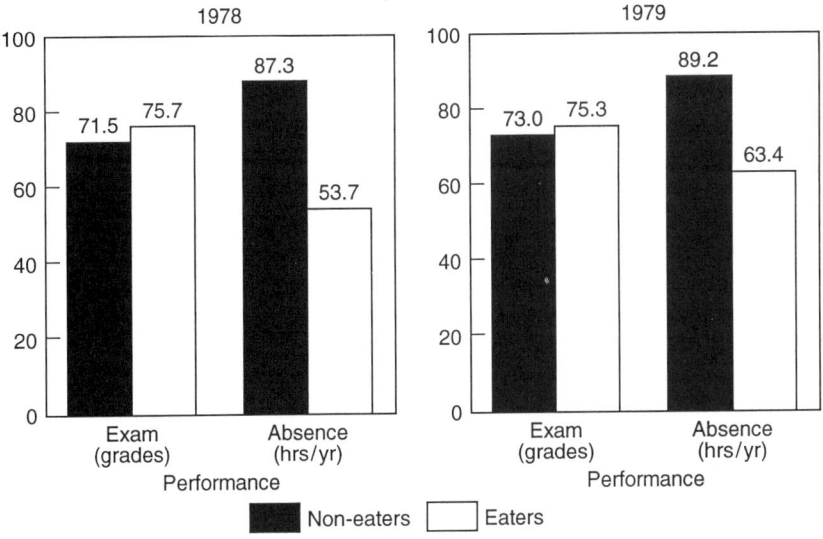

Figure 3.1 Breakfast and its effects on performance of students. (From Kagawa *et al.*, 1980).

findings have been reported among lower income schoolchildren in the US (Meyers *et al.*, 1989) and Canada (Nadeau, 1993).

In recognition that hungry children may be at risk of health problems, and may have difficulty learning, a School Breakfast Program was set up in the US in 1966 and permanently authorized in 1975 by the Federal Government. This is both a nutrition and an education programme, available nationwide, to all students in schools offering the programme. A recent study found that participation in the US School Breakfast Program was positively associated with scores on standardized achievement tests and negatively associated with tardiness and absence rates (Meyers *et al.*, 1988). Similarly, breakfast has also been shown to improve memory, vocabulary testing and rates of attendance in the Peruvian School Breakfast Program (Cueto, Jacoby and Pollitt, 1995; Jacoby, Cueto and Pollitt, 1995).

Perhaps another, less obvious, benefit of the breakfast meal is that it encourages the family to sit down together. Because of work constraints of parents, it may be the only opportunity for the family to eat together. Such an opportunity is likely to benefit the emotional and social wellbeing of school age children.

3.7 Does it matter what we eat?

Health professionals currently recommend that children and adults alike should consume a diet low in fat and high in carbohydrate and fibre (Department of Health, 1992). Research has indicated that a high carbohydrate breakfast can make a major contribution to a reduced fat intake for the entire day (Crawley, 1993; Sommerville and Reagan, 1993; Gibson and O'Sullivan, 1995). There are many studies that demonstrate that breakfasts containing cereals are associated with better diets (Morgan, Zabik and Leveille, 1981; Morgan, Zabik and Stampley, 1986a,b; Zabik, 1987).

Children aged 5–17 years also tend to have lower fat intakes (Morgan, Zabik and Leveille, 1981; Crawley, 1993; Gibson and O'Sullivan, 1995). The proportion of energy from fat has been found to decrease with increased breakfast cereal consumption.

Breakfast cereals are typically high in carbohydrate and low in fat. Many are also high in fibre. Often they are fortified with a range of micronutrients such as vitamins A, B1, B2, B6, B12, niacin, folic acid, D, C and minerals such as iron. When fortified, breakfast cereals have been found to be a major source of important micronutrients in children all over the world. For example in Scotland, fortified breakfast cereals are important sources of vitamins B1, B2 and niacin in the diet of 2–5-year-olds (Payne and Belton, 1992).

British children aged 7–8 years, have been found to have higher micronutrient intakes when breakfast cereals are regularly eaten (Ruxton et al., 1993). British 10–15-year-olds who regularly eat breakfast cereals have higher intakes of vitamins B1, B2, B6 and niacin (Gibson and O'Sullivan, 1995). Higher intakes of vitamins B1, B2, niacin, B6, folic acid and D have been found in breakfast cereal-eating teenagers compared to those who did not eat cereals (Crawley, 1993). Indeed, in this study, a significant proportion of teenagers who did not eat cereal failed to achieve the recommended nutrient intakes of vitamins B1, B6 and folic acid (see Fig. 3.2).

Similarly in Mexico, the diets of frequent breakfast cereal consumers are higher in vitamins A, B6, niacin and zinc (Chavez, 1994).

Recent research has also demonstrated that a low fat and high carbohydrate breakfast is physiologically important in that it primes the body's metabolism for handling subsequent meals. That is, it helps the body to be better able to handle a fatty lunch (Frape et al., 1994). The same research suggests that cereal breakfast eaters have lower risk factors for heart disease and suggests this may also be so for diabetes. Regular breakfast skipping per se may also have long-term adverse metabolic effects such as impaired glucose tolerance and, as already mentioned,

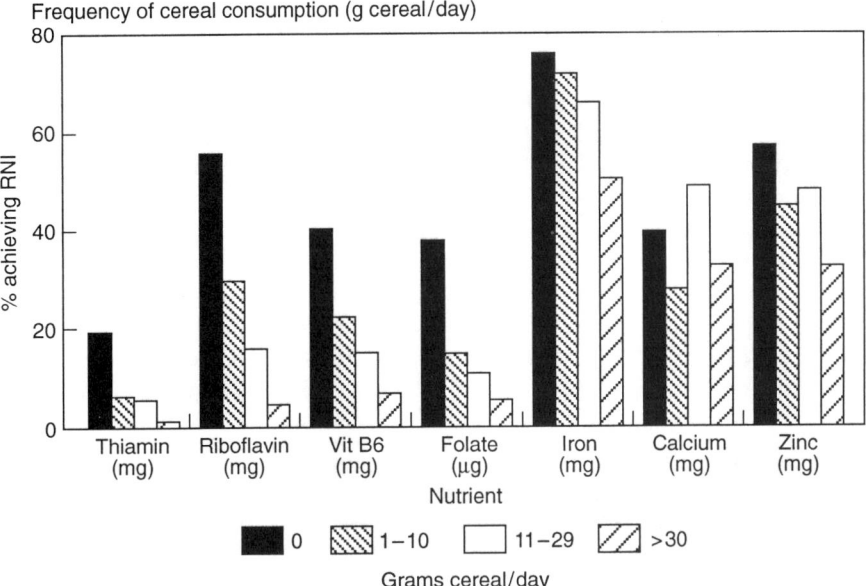

Frequency of cereal consumption (g cereal/day)

Figure 3.2 Percentage of male respondents who did not achieve the RNI for selected micronutrients at four frequencies of breakfast cereal consumption. (From Crawley, 1993).

elevated plasma cholesterol which are risk factors for diabetes and heart disease (Fabry and Tepperman, 1970; Adams and Morgan, 1981).

Another benefit of breakfast cereals is that they encourage the consumption of milk, an important source of protein and calcium for children.

However, breakfast cereals are not without criticism. The sodium and sugar contents are often challenged as being unnecessarily high (pers. comm.). However, put in context, these ingredients which are added to improve the taste and texture of cereals, are not deleterious to health.

Sodium, which comes mainly from salt in the diet, is a nutrient essential for healthy muscle and nerve activity (Department of Health, 1991). Together with potassium, it plays a vital role in controlling the movement of water within the body. Eating too much sodium has been linked with high blood pressure, however there is still considerable disagreement among the experts as to how strong this link is (Swales, 1988; Department of Health, 1991). Other risk factors for hypertension under our personal control, which are probably more significant than sodium intake, are being overweight, smoking, drinking too much alcohol and low physical activity (INTERSALT, 1988; Department of Health, 1991). Breakfast cereals typically provide about 5% of sodium intake (MAFF, 1994).

Sugars, together with other fementable carbohydrates, can play a part in the development of tooth decay. For this to happen, the teeth must regularly come into contact with sugary foods at frequent intervals during the day (the overall amount of sugar is less important). However, this is not the entire story by any means. For example, the rate of dental decay has been falling dramatically throughout the world, but during the same period, the average intake of sugar has hardly changed in most countries (WHO, 1985; Anderson, 1989; Department of Health, 1990; König, 1990). This is thought to be mainly due to the introduction of fluoridation, either via toothpastes or water supplies (Naylor and Murray, 1983; Department of Health, 1990) and better dental hygiene. In the case of breakfast cereals, which in countries like the UK provide less than 5% of the total sugar in the diet (MAFF, 1994), studies show that children who regularly eat the pre-sweetened type, in general have the same standard of dental health as children eating other forms of breakfast cereals (Glass and Fleisch, 1974; Rowe, Anderson and Wanninger, 1974; Finn and Jamison, 1980). This is probably because the sugar from the cereal is quickly removed from the mouth by the milk normally eaten with cereals. Also, breakfast is a once-a-day event, which means the frequency factor is about as low as you can get!

3.8 Can breakfast prevent nutritional problems?

Nutritional problems are common in children from all countries. In developed areas like Europe and the US, iron deficiency anaemia and constipation are still common, while in the developing areas like Latin America, iron, zinc and vitamins A and C deficiencies are prevalent (Bengoa, 1988).

   Poor iron intake is still a major problem for children across the globe. Breakfast cereals fortified with iron are a major contributor to overall iron intake in countries such as the UK (Department of Health, 1989). Anaemia leaves children not only tired, pale and lethargic, but can also affect children's academic development (Department of Health, 1991). According to British data, breakfast cereals are an important source of iron in 2–5-year-olds' diets (Payne and Belton, 1993), in fact it can provide 10% of the total dietary iron intake. Iron intakes are higher in breakfast cereal eaters, at breakfast (Morgan, Zabik and Stampley, 1986a,b; Crawley, 1993) and throughout the day compared to non-cereal eaters (Chao et al., 1984; Meyers et al., 1992). Young girls may be particularly at risk; iron deficiency anaemia is not uncommon in adolescent girls who are menstruating (Department of Health, 1991). Young girls may be at risk of vitamin deficiencies also, because although still growing, they may begin to restrict their dietary intakes. Dieting amongst this age group which is likely to be common, could lead to deficiencies (Mareschi, 1991).

   Another important nutrient is folic acid, a B vitamin. Recent evidence has demonstrated that spina bifida births can be reduced by over 70% if women take adequate folic acid before they get pregnant (Department of Health, 1992). Over 50% of pregnancies are not planned. Therefore, it is important to ensure that women and, indeed, all women of childbearing age, get adequate folic acid in their diets. According to a recent UK Report by the Department of Health (1992), breakfast cereals fortified with folic acid are an excellent source of this vitamin.

Constipation is a very common and relatively unappreciated problem in children. It has been estimated that as many as 44% of young children may be constipated (Holt *et al.*, 1992). Breakfast cereals made from wheat bran, oat bran, psyllium or whole grain are high in fibre and can help to alleviate, and even prevent, constipation. Indeed, wheat bran, a common ingredient in breakfast cereals is more effective than fruit and vegetables in accelerating transit time and, therefore, preventing constipation because it is a concentrated source of insoluble fibre.

Obesity is a serious and growing international health problem among adults in the western world (Seidell, 1992). Although there are no official figures for children, the incidence of obesity is thought to be high. Obesity is a major risk factor for chronic illnesses such as diabetes, heart disease and even some cancers (Department of Health, 1991). Breakfast can play an important role in the prevention of obesity. A French study conducted in 7–12-year-olds found that obese children tend to eat a smaller breakfast and a larger dinner than normal and thin children (Bellisle, 1990). Regular consumption of breakfast cereals is associated with less body fat in children (Gibson and O'Sullivan, 1995). They may also be foods that can help reduce weight, since their consumption is associated with lower dietary fat intakes (Crawley, 1993; Sommerville and Reagan, 1993; Gibson and O'Sullivan, 1995).

Faddy or picky eaters are sometimes a worrying problem for parents of young children. Fortified breakfast cereals are good sources of important vitamins and minerals (Payne and Belton, 1993). Because the range is wide in terms of taste, texture and shape they are appealing to children. They are a good way of encouraging the breakfast habit, while being a good nutritious food.

Since breakfast cereals encourage milk consumption and may be fortified with both calcium and vitamin D, intakes of these nutrients are often higher in children who eat them. Poor intakes of calcium and vitamin D have been reported (Department of Health, 1989). Adequate levels are important for healthy bones and teeth.

3.9 Can the breakfast habit prevent future illness?

There has been much interest in childhood origins of future illness. It is accepted that heart disease risk factors develop in childhood and that obesity in youth usually continues into adulthood (Boreham *et al.*, 1993).

As already discussed, fortified breakfast cereal consumption may help to prevent spina bifida births (Department of Health, 1992). Recent evidence also suggests that vitamins associated with regular consumption of breakfast cereals are important for healthy babies. Indeed, an analysis carried out at a London clinic (Doyle *et al.*, 1990), found that breakfast cereal consumption was associated with fewer low birthweight infants.

The most common chronic illnesses are cancers and heart disease. Diet has long been linked to their development (Department of Health, 1991). High fat intakes, obesity and also poor fibre intakes are risk factors.

A recent study found that the consumption of a high fibre breakfast cereal may help to decrease energy intake at lunch (Delargy *et al.*, 1995). These results are intriguing since a reduced energy lunch, combined with a breakfast that is somewhat reduced in calories due to the food energy 'dilution' effect of fibre, could potentially result in weight loss if maintained over time.

Colorectal cancer is a major cause of death across the world and, in some countries (Gori, 1978) the incidence is increasing. Typically, the rates are low in populations that traditionally consume diets rich in fibre (Burkitt, 1971). Studies using laboratory animals are now showing that grain fibre, particularly wheat bran, may be more strongly associated with lower risk of this cancer than other fibre sources (Shivapurkar, Tang and Alabaster, 1992). Recent studies suggest that wheat bran may be beneficial in also reducing cancer risk in humans (DeCosse, Miller and Lesser, 1989; Alberts *et al.*, 1990). The accumulating science suggests not only colon cancer, but also breast cancer risk may be reduced by wheat bran (Rose, 1990).

Some breakfast cereals contain wheat bran. With consumers looking for easy ways to make their diets more healthy, high fibre breakfast cereals offer a practical, quick and tasty answer.

Two of the major risk factors for heart disease are a high plasma cholesterol level and being overweight (Department of Health, 1991). Breakfast cereal eaters tend to have lower levels of cholesterol in their plasma (Stanton and Keast, 1989; Resnicow, 1991) and also tend to be slimmer (Gibson and O'Sullivan, 1995). Recent and quickly developing science suggests a new risk factor. A high level of an amino acid called homocysteine in the blood is now thought to be an even stronger risk factor for heart disease than plasma cholesterol (Masser, Lloyd and Potter, 1994). It seems that vitamins like folic acid, B6 and B12 can reduce these levels and, therefore, may help prevent the risk of heart disease. This is clearly an important benefit of fortification and also another reason why fortified breakfast cereals can affect health positively.

3.10   Start the day right!

To summarize, there is an immense wealth of information on breakfast. Eating breakfast not only improves overall dietary intake and nutritional status but, according to the growing science, also contributes to learning success. Eating breakfast may help to control body weight and possibly lower the risk of heart disease. Recent evidence also suggests that high fibre breakfast cereals may help to reduce risk of cancers that are associated with poor fibre intakes. In fact, breakfast may also positively affect mood.

Because dietary habits in adulthood are developed during childhood, it is important that children be encouraged to eat a healthy diet. Adolescence is a transitional stage when the structure of food habits may become relaxed. Healthy eating practices must be formed by this age.

Current dietary guidelines (Department of Health, 1991) focus on decreased consumption of fat with increased emphasis on foods supplying carbohydrate and fibre. Easy to prepare carbohydrate foods such as breakfast cereals which are low in fat can help constitute a balanced diet that is consistent with nutrition recommendations.

Starting the day right takes on a whole new meaning when we consider the short-term benefits and also the longer-term benefits of eating a cereal breakfast.

References

Adams, C.A. and Morgan, K.J. (1981), Periodicity of eating: Implications of human food consumption. *Nutrition Res.*, **1**, 525–50.

Alberts, D.S., Einspahr, J., Rees-McGee S. *et al.* (1990), Effects of Dietary Wheat Bran Fiber on Rectal Epithelial Cell Proliferation in Patients With Resection for Colorectal Cancers. *J. Natl. Cancer Inst.*, **82**, 1280–5.

Anderson, R.J. (1989), The changes in dental caries experience of 12 year old schoolchildren in two Somerset schools. A review after an interval of 25 years. *Br. Dental J.*, **167**, 312–4.

Arnold, L.M. *et al.* (1993), Effect of isoenergetic intake of three to nine meals on plasma lipoproteins and glucose metabolism. *Amer. J. Clinical Nutrition*, **57**, 446–51.

Bellisle, F. (1990), Obésité de l'enfant: comportement alimentaire et variables socioculturelles. *Cah. Nut. Diét XXV*, **4**.

Bengoa, J.M. *et al.* (1988), *Nutritional goals and dietary guidelines for Latin America. Bases for Development.* Cavendes Foundation, Caracas, Venezuela.

Boreham, C., Savage, J.M., Primrose, D., *et al.* (1993), Coronary risk factors in schoolchildren. *Archives of Diseases in Childhood*, **68**, 182–6.

Burkitt, D.P. (1971). Epidemiology of cancer of the colon and rectum. *Cancer*, **28**, 3–13.

Chao, E.S.M., Anderson, G.H., Thompson, G.W. *et al.* (1984), A longitudinal study of the dietary changes of a sample of Ontario children II: Food intake. *J. Canadian Dietetic Assoc.*, **45**(2), 112–20.

Chavez, A. (1994), *Urban survey on diet and cereal consumption in Mexico City, Mexico.*

Department of Community Nutrition, National Institute of Nutrition, Mexico City, Mexico.

Crawley, H. (1993), *J. Human Nutrition and Dietetics*.

Cueto, S., Jacoby, E. and Pollitt, E. (1995), Breakfast prevents delays of memory function. *FASEB Atlanta, Georgia*, Abstract No. 2801.

DeCosse, J.J., Miller, H.H. and Lesser, M.L. (1989), Effect of wheat fibre and vitamins C and E on rectal polyps in patients with familial adenomatous polyps. *J. Natl. Cancer Inst.*, **81**, 1290–7.

Delargy, H.J., Burley, V.J., O'Sullivan, K. *et al.* (1995), Effects of soluble and insoluble fibre at breakfast on 24 hour pattern of dietary intake and satiety. *European J. Clinical Nutrition* (in press).

Department of Health (1989), *The diets of British schoolchildren*, No. 36. HMSO, London.

Department of Health (1990), The annual report of the Chief Medical Officer of the Department of Health. Dental Health. *Br. Dental J.*, **168**, 76–77.

Department of Health (1991), *COMA. Dietary Reference Values for food energy and nutrients for the United Kingdom*, No. 41. HMSO, London.

Department of Health (1992), *Folic acid and the prevention of neural tube defects.* Department of Health, UK.

Doyle, W., Crawford, M.A., Wynn, A.H.A. and Wynn, S.W. (1990), The association between maternal diet and birth dimensions. *J. Nutritional Medicine*, **1**, 9–17.

Edelstein, S.L. *et al.* (1992), Increased meal frequency associated with decreased cholesterol concentrations; Rancho Bernardo CA, 1984–1987. *Amer. J. Clinical Nutrition*, **55**, 664–9.

Fabry, P. and Tepperman, J. (1970), Meal frequency – A possible factor in human pathology. *Amer. J. Clinical Nutrition*, **23**, 1059–68.

Finn, S.B. and Jamison, H. (1980), The relative effects of three dietary supplements on dental caries. *ACDC J. Dentistry for Children*, **47**, 33.

Frape, D.L., Williams, N.R., Pickersgill, J. *et al.* (1994), The postprandial response to high fat, low carbohydrate and low fat high carbohydrate meals providing the same amounts of energy in subjects at risk of heart disease. *Proceedings of the nutrition society*, 221A.

Gibson, S. and O'Sullivan, K.R. (1995), Breakfast cereal consumption patterns and nutrient intakes of British schoolchildren. *J. Roy. Soc. Health*.

Glass, R.L. and Fleisch, S. (1974), Dental caries incidence and the consumption of ready to eat breakfast cereals. *J. Amer. Dental Assoc.*, **88**, 807.

Gori, G.B. (1978), Role of diet and nutrition in cancer cause, prevention and treatment. *Bull. Cancer*, **65**, 115–26.

Guldon, G.S., Tao, W., Fung, I. and Leung, F. (1994). Breakfast eating habits of children in Hong Kong. Presented at the 'Childhood Nutrition Seminar', Hong Kong, June 1994.

Health and Welfare Canada (1977), *Nutrition Canada: Food Consumption Patterns Report.* O'Hawa: Bureau of Nutritional Sciences, Health Protection Branch.

Holt, M.A., Sali, A., Kucherhan, M. and Thompson, G. (1992), Constipation and diet in primary school children. *J. Nutritional Med.*, **3**, 257–62.

Howden, J.A., Chong, Y.H., Leung, S.F. *et al.* (1993), Breakfast practices in the Asian Region. *Asia Pacific J. Clinical Nutrition*, **2**, 77–84.

INTERSALT Co-operative Research Group (1988), Intersalt: an international study of electrolyte excretion and blood pressure. Results for 24 hour urinary sodium and potassium excretion. *Br. Medical J.*, **297**, 319–28.

Jacoby, E., Cueto, S. and Pollitt, E. (1995), Dietary and cognitive effects of a school breakfast program among Andean children of Peru. *FASEB Atlanta, Georgia*, Abstract No. 2802.

Jenkins, D.J.A. *et al.* (1989). Nibbling versus gorging: metabolic advantages of increased meal frequency. *New England J. Medicine*, **321**, 929–34.

Kagawa, Y., Nishimura, K., Sato, J. *et al.* (1980), Omission of breakfast and its effects on the nutritional intake, serum lipids and examination grades of dormitory students. *Japanese J. Nutrition*, **38**(6), 288–94.

König, K.G. (1990), Changes in the prevalence of dental caries: how much can be attributed to changes in diet? *Caries Research*, **24** (suppl. 1), 16–18.

MAFF (1994), *National Food Survey 1993*. HMSO, London.

Mareschi, J.P. (1991), Identification of those micronutrients most likely to be insufficient as the result of habitual low energy intake. In *Modern Lifestyles, Lower Energy Intake and Micronutrient Status*, Pietrzik, K. (ed.), Springer-Verlag, London, pp. 45–50.

Martinez, O.B. (1982), Growth and dietary quality of young French Canadian schoolchildren. *J. Canadian Dietetic Assoc.*, **43**, 28–35.

Masser, P.A., Lloyd, M.T. and Potter, J.M. (1994), Importance of elevated plasma homocysteine levels as a risk factor for Atherosclerosis. *Annals of Thoracic Surgery*, **58**, 1240–6.

McGrath, S.A. and Gibney, M.J. (1994), The effects of altered frequency of eating on plasma lipids in free-living healthy males on normal self-selected diets. *European J. Clinical Nutrition*, **48**, 402–7.

Meyers, A.F., Sampson, A., Weitzman, M. and Kayne, H.L. (1988), School breakfast program and school performance. *Amer. J. Diseases in Childhood*, **142**, 398.

Meyers, A.F., Sampson, A.E., Weitzman, M. *et al.* (1989), School breakfast program and school performance. *Amer. J. Diseases of Children*, **143**, 1234–9.

Morgan, K.J., Zabik, M.E. and Leveille, G.A. (1981), The role of breakfast in the nutrient intake of 5 to 12 year old children. *Amer. J. Clinical Nutrition*, **34**, 1418–27.

Morgan, K.J., Zabik, M.E. and Stampley, G.L. (1986), The role of breakfast in diet adequacy of the U.S. adult population. *J. Amer. College of Nutrition*, **5**, 551–63.

Morgan, K.J., Zabik, M.E. and Stampley, G.L. (1986), Breakfast consumption patterns of older Americans. *J. Nutrition for the Elderly*, **5**(4), 19–43.

Nadeau, M.H. (1993), Nutrition and academic achievement: the importance of cultural differences. *Rapport*, **8**(2), 4–5.

Naylor, M.N. and Murray, J.J. (1983), *The Prevention of Dental Disease*, Oxford University Press, Oxford.

Nicklas, T.A. *et al.* (1993). Breakfast consumption affects adequacy of total daily intake in children. *J. Amer. Dietetic Association*, **93**(8), 886–91.

O'Flaherty, L. and Gibney, M.J. (1994), The effect of very low-, moderate- and high-fat snacks on postprandial reverse cholesterol transport in healthy volunteers. *Proceedings of the Nutrition Society* (in press).

Payne, J.A. and Belton, N.R. (?), Nutrient intake and growth in pre-school children. II. Intake of mineral and vitamins. *J. Human Nutrition and Dietetics*, **5**, 299–304.

Pollitt, E., Leibel, R.L. and Greenfield, D. (1981), Brief fasting, stress and cognition in children. *Amer. J. Clinical Nutrition*, **34**, 1526–33.

Pollitt, E., Lewis, N.L., Garza, C. and Shulman, R.J. (1982/3), Fasting and cognitive function. *J. Psychiatric Res.*, **17**, 167–74.

Resnicow, K. (1991), The relationship between breakfast habits and plasma cholesterol levels in schoolchildren. *J. School Health*, **61**(2), 81–5.

Rose, D.P. (1990), Dietary fibre and breast cancer. Nutrition and Cancer. *Nutrition and Cancer*, **13**, 1&2, 1–8.

Rowe, N.H., Anderson, R.H. and Wanninger, L.A. (1974), Effects of ready to eat breakfast cereals on dental caries experience in adolescent children: a three year study. *J. Dental Res.*, **53**, 33.

Ruxton, C.H.S. (1993). Breakfast Habits in Children. *Nutrition & Food Science*, **4**, 17–20.

Seidell, J. (1992), The prevalence of obesity in Europe. *Proceedings of the First European Dietary Fibre Seminar, Copenhagen*.

Shivapurkar, N., Tang Z. and Alabaster, A. (1992), The effect of high risk and low risk diets on aberrant crypt, colonic tumour formation in Fischer-344 rats. *Carcinogenesis*, **13**, 5, 887–90.

Simeon, D. and Grantham-McGregor, S. (1987), Cognitive Function, Undernutrition and Missed Breakfast. *The Lancet*, September 26, 737–8.

Skinner, J.D., Salvetti, N.N., Ezell, J.M. *et al.* (1985), Appalachian adolescents' eating patterns and nutrient intakes. *J. Amer. Dietetic Assoc.*, **85**, 1093–9.

Stanton, J.L. and Keast, D.R. (1989), Serum cholesterol, fat intake, and breakfast consumption in the United States adult population. *J. Amer. College of Nutrition*, **8**(6), 226.1–226.6.

Swales, J.D. (1988), Salt saga continued. *Br. Medical J.*, **297**, 307–8.

WHO (1985), Changing patterns of oral health and implicated for oral health manpower. Part 1. *International Dental J.*, **35**, 235–51.

Yap, M. (1993), Dietary habits in Singapore schoolchildren. Presented at the 'Childhood Nutrition Seminar', Singapore, May 1993.

Zabik, M.E. (1987), Impact of ready-to-eat cereal consumption on nutrient intake.

4 Why tastes change

S. LANG

4.1 Introduction

The nature of childhood is changing. The influences on children in the mid-1990s can scarcely be compared with the influences on their parents, when they were children. The word 'computer' was hardly in the dictionaries. The concept that 'we are what we eat' was a novelty. The family was a hierarchical structure, and not today's more democratic unit. With each child is a new generation of parents, with new sets of circumstances. Each generation adapts, of course, to the pace of the times, and children react to those changes in their particular ways, and quickly.

There are three particular areas of change.

- Children have ideas, knowledge and skills their parents did not have, through education and media. Their interests can be as wide as those of any adult.
- The children of today grow up with different behaviour patterns, and interests. They schedule their time, their goals and role models to the financial, working and living necessities of their parent(s), and to the environment of the times.
- Children have significant influence on all aspects of family expenditure – food, clothes and leisure – and responsibility for their own time management.

The Victorian epithet that children should be seen and not heard was the discipline for millions until perhaps the advent of World War Two. Children should remain in their place in an ordered world, as society developed within parameters of familiar – if not comfortable – structures. Within a generation, there has been irrevocable change in the role of children in society. Children are consumers – in every commercial sense – because they are exposed to adult influences. Families expect contribution and comment to the family decision-making process. Children share and contribute to their world in scaled-down adult ways.

Until the early 1970s, for example, any appraisal of children and their food, and then its marketing and advertising was through the eyes of Mum. Breakfast cereal or confectionery marketing trod the challenging but fine line of a dual appeal to children – the consumers, and to adults – the buyers. For children, the appeal would be the fun, the sound, the colours, the enjoyment of the product. For parents the appeal was that cereal was an important part of the diet, or that chocolate was good for kids.

But change is a-foot. In the UK, 85% of men and 71% of women (up from 57% in 1971) are employed full-time or part-time. There has been a shift in parents' (even grandparents') attitudes and patterns of behaviour. Families expect more responsibility and decision-making from children: a far cry, for children, indeed, from being seen and not heard!

However, what has changed? It is not children themselves who have changed, of course, but parents and their attitudes. Parents have seen enormous developments in society and work during the 1980s and 1990s. Through society, family and media, children absorb the information that gives knowledge, that leads to power. The effect on children? Children have influence in family life. The era, almost, of 'child empowerment'.

This chapter looks at how society's changes are affecting families and children and food. We shall be looking at children and how they influence food. It will show that:

● the changes in the structure of adult society have prompted different parental demands and expectations of their children;
● there has been significant change in the foods that families eat;
● children are an influential force in society.

4.2   Key changes in society, family and food

Our society is changing a lot. Parents swim with that change, and so do their children. Society and families change by adapting to innovation and circumstance. Developments in society and technology over the last 15 years have added new dynamism to the impact on children. These developments have been dramatic, and have opened new dimensions in the structure of the traditional family.

We are seeing, now, important breaks in that structure. The trends in employment, household size or population ageing, for example, are not new. But the combination of many of the trends, combined with the contribution of new technology, have changed families, how children are brought up, what families eat, and how they cook.

4.2.1   Employment changes of women

Women are a large percentage of the work force, and this percentage is likely to grow: in 1993, 50% of the work force was female.

At the same time, part-time workers as a percentage of the labour force have increased. In 1993, as Table 4.1 shows, 28.3% of the work force was

Table 4.1 The changing composition of the labour force

Year	Part-time work %	Self-employed work %	Men in employment ($\times 10^3$)	Women in employment ($\times 10^3$)	Total in employment ($\times 10^3$)
1972	16	9.6	13 620	8 517	22 137
1982	NA	11.1	12 038	9 104	21 143
1987	23.2	14.3	11 876	10 153	22 028
1993	28.3	15	10 881	10 695	21 576

Source: Henley Centre, Planning for Social Change (1995/96).

made up of part-time workers, but what is critical is that the major proportion (81%) of these people are women.

Thus, in the space of a generation, from the mid-1960s, we have seen the work force move from unmistakably male – with the attitudes and behaviour patterns that engenders – to a near equal balance of men and women. The norm for the children of today is that both parents work.

In this, as in other areas, traditions are changing and becoming established in different ways. The home has become more the motel, less the castle, as all the family members move in and out from school or work. It is hardly surprising therefore to find that:

- 57% of all parents with children under 15 agree that they try to spend as little time as possible preparing and cooking food;
- 57% of women, and 48% of men, agree that convenience foods make their lives easier;
- only a quarter (26%) of men and women really enjoy food shopping.

New traditions and patterns are being set, which give greater freedom and responsibility to children at a younger age: the adage that quality of parental input is more important than quantity *per se* is true; while other families or grandparents or neighbours or television take on the role of minding and guiding children as they grow up. The resurgence of the nanny is hardly surprising, and the rise in 'teleworking' lets more and more parents work at home and accommodate school-aged children at the same time.

4.2.2 Time-use: changes in adult cooking patterns

Does this apparent decline in parents' contribution mean that parents feel they are giving less to their children? Do they feel they are performing less than what they expect of themselves? To an extent it does. The basic social and 'mother-provider' instincts are far from dormant (Table 4.2).

If they had more time, over half of all women (54%) would try out new

Table 4.2 If you had more time, which of these would you do?

	%
Try out new recipes more often	54
Buy fresh food more regularly	47
Cook healthier food more often	47
Invite people to my house for a meal more often	46
Cook from basic ingredients more often	40
Plan meals more carefully	34
Visit more than one food store per week	21
Write a shopping list for every grocery shopping trip	20

Source: Henley Centre Omnibus Survey (1993).

recipes more often; and just under half (47%) would buy fresh food more regularly and cook healthier food more often; while 40% would cook from basic ingredients more often. Mothers have an underlying sense of guilt that they should contribute more to their families.

Changing employment and family living patterns have altered the way families think about food and prepare and cook it – in some cases almost the concept of the meal itself. Meals are no longer the family meeting time, particularly as children become more self-sufficient. Nowadays, 64% of children prefer eating snacks to main meals, and 35% of children over five do not always eat the same food as their parent. Sitting down together is not the norm.

However, this is not to suggest that we are becoming a nation whose future generations will be brought up by guilt-ridden mothers, who have so little spare time after working hard that they are not feeding their children properly. We always feel we can do better when it comes to our families and children: an endemic feature of motherhood is not to feel complacent or self-satisfied! Indeed, although the mums of the nation may feel guilty, the facts are that we are now eating more healthily than a generation ago.

4.2.3   Are we eating more healthily?

Since 1965, British eating habits in general have altered (Table 4.3). We eat more rice, breakfast cereals, poultry and fruit, and less fat and red meat: key ingredients in the guidelines for eating more healthily.

On the whole, for health reasons, this is an encouraging picture. But we are also eating fewer vegetables and less fish and bread; and – though not in Table 4.3 – we are eating more crisps, chips and chocolate biscuits. Table 4.3 provides an interesting and encouraging picture, because it runs counter to the perceptions that most of us would have of how, in general, our diet is changing.

Table 4.3 How the British diet in general has changed 1965–1992 (% change in food consumption)

	% change
Rice	138
Breakfast cereals	137
Poultry	126
Coffee	43
Bananas	42
Cheese	25
Fruit	9
Other cereals	5
Total meat	–11
Vegetables	–12
Biscuits	–14
Fish	–14
Milk and cream	–25
Total fats	–27
Bread	–35
Beef	–38
Sausages	–42
Cakes and pastries	–45
Tea	–48
Eggs	–57
Sugar and preserves	–66
Butter	–76

Source: National Food Surveys/Henley Centre.

Advertising, articles in the media, and the sight or sound of nagging and pestering children would give us the impression that healthy eating is a long way off. Undoubtedly better information about balancing a diet has contributed to this improvement, though we cannot at the same time overlook the strides taken by the food industry. Today's large supermarkets will stock between 20 000 and 40 000 different products. Supermarkets have all but eliminated seasonality in what we can buy, but at the expense of freshness. Supermarkets have made it easier for households to establish patterns of shopping and food preparation, as well as making sensible eating easier and less of a chore.

4.2.4 Food preparation

We prepare our food more healthily nowadays. The traditional methods of frying, roasting and baking are now used less. There are a number of reasons for the changes in the way we prepare our food.

1. People have less discretionary time: women are spending less time cooking, and though men cook more than they used to, they still fall

short of women. People spend on average 7.49 hours a week cooking – just over an hour a day – and forecasts are that this will decline to less than an hour a day, as people search out convenient cooking options.

2. There is wide knowledge that roasting, baking and frying – with fat – is less healthy than other methods. In 1992, 40% of our food energy was accounted for by fat (16% of which is saturated fat). This information has led to wide exhortations to cut down on frying. Roasting or baking is used in only 15% of meals and frying in only 11% (Table 4.4).

3. Technology has made available new methods, such as microwaving, which enable us to avoid the less healthy ways of cooking. The chart shows that microwaves were used to prepare food in 7% of meal occasions – and we anticipate this will increase as convenience foods and other food types carry microwave instructions. There is wide ignorance and, perhaps, a fear of technology about microwaving. In 1988, 33% of households owned a microwave, and by 1995 this had all but doubled (to 64%), yet in only 7% of meal occasions was the microwave used.

4. Microwaves also put a safe cooking method within the easy reach of people with no fear of technology – children. Twenty or thirty years ago, children had none of the cooking opportunities of today. The very idea of children preparing a meal was ridiculous, as well as parents being unwilling, and the means and procedures too prolonged.

5. An additional factor changing the way we prepare our foods is the development of convenience foods, requiring healthier preparation methods. Convenience foods now account for over 5% of the family food budget – a figure we anticipate will increase as people have less time to spend on cooking.

Table 4.4 Preparation techniques (% of meal occasions where technique was used)

	% of occasions used
Uncooked/cold	87
Grilled/toasted	29
Boiled/steamed	23
Heated	18
Roasted/baked	15
Fried	11
Microwaved	7
Stewed	2

More than one method of preparation can be used in one meal.
Source: Taylor Nelson Food Panel (1992).

4.2.5   Changes in family eating

The presence of children in the home fosters a responsible attitude to family feeding, and creates additional changes in eating patterns.

There is no question but that we are a meat-eating nation. If we look at what a typical family of 2 adults and 2 children eats, a quarter of their budget (24.76%) goes on meat, with a further 16% going towards milk, cream and cheese (Table 4.5). The lion's share, however, is spent on cereals, vegetables and fruit, which between them account for almost 42% of the family's budget.

What is perhaps more revealing is the way that the family's diet has been changing over the years. The indirect influence of children has been quite marked. More noticeably, there have been dramatic declines of between 24% and 49% over this 12-year time span in the proportion of its budget that the family spends on fats, beverages, eggs and sugar and preserves. No less dramatic, because of the significant effect it is having on the industry, is the 17% decline in the spend on meat – all the decline being in red meat.

By contrast families are now spending more of their food budget on cereals (up by 18%), vegetables (24%), fruit (29%) and, to a lesser extent, they are also eating more fish (up by 8%).

So that we can appreciate the full picture, on average £11.01 is spent on food for each person in the family every week. Thus, each person is now eating some £2.73 worth of meat each week, £2.13 of cereals, and £1.61 of vegetables – Table 4.6 gives the full details. Seen in this way, it seems a little surprising, given they are nature's original convenience food, that the average family member only eats one egg a week.

Table 4.5 Family food spend (2 adults and 2 children: 1980 vs. 1992)

	% of weekly food spend	% change 1980 vs. 1992
Meat	24.76	−16.7
Cereals	19.35	17.8
Vegetables	14.66	23.9
Milk and cream	12.45	0.8
Fruit	7.97	28.8
Miscellaneous	5.17	33.9
Fish	4.41	7.8
Cheese	3.72	6
Fats	2.67	−35.4
Beverages	2.49	−24.3
Eggs	1.32	−48.8
Sugar and preserves	1.04	−49.3

Source: National Food Survey.

Table 4.6 How much do people spend on food?

	Average weekly spend per person (pence)
Meat	273
Cereals	213
Vegetables	161
Milk and cream	137
Fruit	88
Fish	49
Cheese	41
Fats	29
Beverages	27
Eggs	15
Sugar and preserves	12
Total spend per person	£11.01

Source: Asda/Henley Centre.

4.2.6   The indirect influence of children on food

What families eat has changed significantly, and quite quickly – these changes are over only a 12-year time span. In essence, the changes are that families with children are spending less of their budget on meat and more on fruit, vegetables, cereals and fish. Here, clearly, we need to explore the factors which are accounting for this trend in family diets.

- There is a trend towards eating more vegetarian foods, taking diets away from red meat. Our work shows that 11% of girls aged 13–15 avoid eating red meat. This is a high figure in itself: indeed there are indications that, of the next generation of mothers (girls who are 10- to 15-years-old now), some 20% will avoid meat-based dishes. Concerns about BSE (Bovine Spongiform Encephalopathy), the colloquially termed 'mad cow disease', are never far from the headlines and have contributed to the decline in red meat, as they have to the increase in white meat consumption.

- Being vegetarian – or avoiding red meat – is no longer a quirk. Although in a minority, 3% of children say they eat healthy food because they are vegetarian (Table 4.8). To avoid red meat is no longer considered abnormal; to be vegetarian is no longer cranky. Children are keen to become and remain healthy. They take their sport seriously: '*mens sana in corpore sano*'.

- Indeed, children show every indication of being more health-aware than adults, as Table 4.7 shows. They have a more practical understanding of health – an appraisal of exercise and health fads, for example – than adults, in general. This is likely to reflect what children learn from school

Table 4.7 Children are more health aware than adults

| % agreeing that . . . | Adults (total) | Children | |
		9–10 years	11–12 years
Brown sugar is better than white	39	40	54
Can't do much to be healthier: illness is chance	32	30	27
Daily exercise means you can eat whatever you want	29	19	18
Health fads are nonsense	38	19	27

Source: Asda/Henley Centre – Health and Diet Survey (1993).

Table 4.8 Why do children eat healthy food?

Unprompted responses	% eat healthy food because . . .
My mum/dad makes me eat healthy food	50
I enjoy eating healthy foods	23
Healthy food is good for you	9
I am worried about my weight/on a diet	7
They tell us to eat healthily at school	7
To keep fit for sport	5
I'm vegetarian	3

Base: Children claiming to eat healthy food.
Source: Asda/Henley Centre – Children, Health and Diet Survey (1993).

and their closer active and passive involvement with sport. This greater awareness is endorsed by what children themselves say when asked why they eat healthy food (Table 4.8).

Certainly there is parental pressure – 50% say their mum or dad makes them eat healthy food. Yet what is revealing (a further sign that children and adults are exposed to the same influences) is that fully 40% of children eat healthy food out of a positive commitment to their wellbeing: they enjoy them, or see them as healthy, or in other cases children like to keep fit for sport, or, as we have discussed, in 3% of cases claim to be vegetarians. They show greater understanding that 'we are what we eat' than children of a generation ago. This is the influence, clearly, of TV programmes, of the greater feeding and family responsibilities asked of them, and of manufacturers' openness about nutritional information.

• Information – which in many cases children will glean at the same time as their parents, from a TV programme, for example – has led to children's growing awareness and understanding of the healthiness of foods. Table 4.9 shows that there have been significant changes among children in their awareness of the types of food that they think are healthy.

Table 4.9 Would you say the following types of food are healthy?

	1986 (%)	1993 (%)	% difference 1986–1993
Bananas	94	92	–2
Skimmed milk	82	85	4
Cheese	84	68	–19
Yoghurt	87	80	–8
Butter	44	22	–50
Ice cream	25	16	–36
Biscuits	22	9	–59
Margarine	21	27	6
Cola drinks	15	8	–47
Hamburgers	27	8	–70

Base = children 9–15 years.
Source: Henley Centre Children and Health survey (1993), Asda/ Henley Centre – Children, Health and Diet Survey.

A lot of these changes are quite marked, especially since they have taken hold over only a 7-year time span. People argue that children are forming incomplete views about certain foods. Are hamburgers healthy? Only 8% of children see them as healthy.

Children's views change quickly. Their influence on what the family eats is affected by what parents deem to be good for their children, and quite directly by what children themselves believe to be healthy. This is important. Word-of-mouth in the playground is one of the most potent advertising media. But so is the added sugar nutritional information printed on yoghurt packs or cola cans, or the national concerns about cholesterol.

The concern about health is no longer the high ground of grown-ups alone. It is practical and important knowledge for children: their knowledge turns to potent influence within the family. Adults and children share similar views about foods that are good or bad for you: they all agree that skimmed milk and yoghurt are healthy; and all agree that biscuits, butter, cola and hamburgers are unhealthy foods.

- Further influences on how children learn about food are people and media. We have hinted at the importance of parents, of course, and of TV and school. The respective influences show up in Table 4.10.

Some 40% of 9–14-year-olds look to their doctor for advice on health. Then among 13–14-year-olds, the importance of the doctor declines, and the influence of the teacher grows. Yet, it is parents who become the main source of advice among 13–14-year-olds, while almost 1 in 5 (18%) look to the media – TV and posters or leaflets – for their main advice.

Table 4.10 Sources of information on health: which of the following gives you the most important advice on health?

	All children 9–14 years old	13–14-year-olds
Doctor	40	28
Parents	31	30
School teachers	14	22
TV	6	9
Posters/leaflets	5	9
Friends	3	6

Source: Asda/Henley Centre 'Children and Food' Survey (1993); Henley Centre/Nielsen 'Measures of Health' Survey.

So, we see a picture of children in general looking across a range of sources for advice, in the same way that adults will form their own views from talking to a variety of people. It is of more than passing interest that friends are only a minor source of advice on health. Children are able to differentiate genuine advice from scuttlebutt: a wide network of people and sources keep children well informed.

4.3 The direct influence of children on food

We have discussed a number of the influences that children have on food. A number of factors – such as children in a family, the role of parents, doctors and teachers, health trends away from red meat – create a greater health awareness and lead to different diets among families with children.

However, parents' working patterns are changing. Around one-third of households in the UK have children where it is now the norm that both parents are working full-time or part-time. In the case of the 17% of families (or 5% of households) that are single parent there are responsibilities and time constraints that have altered family patterns. By 1990, only 11% of families conformed to the stereotype of the father going out to work, with the mother remaining at home – though this figure will change as growing numbers of people telework, work part-time and are self-employed.

This upheaval in the traditional family is reflected in the attitudes of women to having children. Table 4.11 shows a near doubling of the proportion of women across all occupational grades who return to work within 8–9 months of giving birth. Infant feeding and maternal bonding are evolving different patterns and responsibilities, and perhaps leading to a re-emergence of the extended family.

Thus the culture of the family is changing. Parents need to earn money;

Table 4.11 Returning to work: % of women who were full-time employees during pregnancy who were in work 8–9 months after the birth

	1979	1988
I or II	32	59
IIIN	11	32
IIIM	19	28
IV	15	29

Source: Policy Studies Institute.

they have careers to follow; their home is the motel, not the castle. This change shifts adults away from parenting as an objective for (emotional) security, to income-earning as the means of **buying** security. Parenting may be one part of life. Adults are not self-centred, but children's growing-up process is changing. Responsibility is given to them earlier, and greater contribution to the family process is expected of them.

- Children mature physically earlier than they used to. Tambrands research has shown that girls start to menstruate on average at 12.5 years old, which is some 4 to 5 years earlier than a century ago.
- Some commentators have suggested that a mark of children in the 1990s is the race towards adulthood. There is pressure on children to perform well from the earliest stages. The attitudes of early adolescents are adult in outlook, as is the way they set their own values: they practise for an imminent future.
- Of course, physical development and mental attitude will always distinguish between adult and child, but the differences are few in some areas. Without the trappings of adulthood, children are more easily able to copy adult behaviour and mimic their attitudes. They talk on the phone like veterans, have grown up with computers as part of life, use a microwave instinctively, programme a VCR with the skill of space programmers, and use e-mail more naturally than writing a shopping list.

The question, then, is less one of asking what children have yet to learn on their way to adulthood, but more one of asking what adults can do that children can't!

Many of the manipulative and understanding skills required of tomorrow's workers in service industries are acquired early in life. Adults know that children are able to manipulate most things, and therefore expect them to. The media shape a world that is common to both, in a way that technically – and attitudinally – was not possible a generation ago.

The outcome is that parents feel more comfortable about this early responsibility, and invite their children's early contribution to domestic life and decision-making. It not only gives children experience, but also frees

Table 4.12 Influence of 5–11-year-olds: how often does the child have any influence on the final decision made?

	% of parents saying influence always or most of the time
Kid's clothing	37
Cinema film	30
Day out destination	23
Choice of video film	21
Food	11
Holiday destination	10
Choice of TV programme	9
Family restaurant	8
Furniture item	6
Parent's clothes	1

Source: 'Kids as Customers' by James U. McNeal (USA); Henley Centre Planning for Social Change programme.

up parents to pursue their own lives and, of course, their own money-making needs.

This affects the 'role' that children have within family units. Indeed, we have found that:

- 75% can microwave;
- 45% of children sometimes use a microwave to prepare themselves a snack or a full meal;
- 38% of children do not eat the same food as their parents;
- 32% say they prepare a meal on their own once a week or more;
- and almost one-third of families do not eat main meals together most of the time.

4.3.1   Influence of 5–11-year-olds

So we are seeing evidence of children's direct contribution to the family, and most tellingly, this contribution rises dramatically as children move into their teens.

Given our earlier discussions, we would expect that the direct influence of even 5–11-year-olds would be quite significant in some areas. This proves to be the case, where we can see (Table 4.12) that over a third (37%) of 5–11-year-olds influence their own choice of clothes most of the time, and 30% have a similar influence over the choice of cinema film. Over a fifth have significant direct influence in choosing where to go for a day out, or which video film to watch. And just over 1 in 10 (11%) call the tune in buying food all or most of the time.

If we take the meals of the day, the range of direct influence by children aged 5–11 is considerable. They have a 10% influence on breakfast cereals, on sandwich fillings and bread, on crisps and confectionery (apart from what they buy independently), biscuits, snacks, vegetables, convenience foods, meats, ice cream, malted drinks and so on. Across a range of brands and products, there is little that children – at even this young age – do not have some impact on.

4.3.2   Influence of 12–16-year-olds

As adolescence fills out the mind and body, parents and society demand greater responsibility of their children. Children aged 12–16 play double the part in choosing their own clothes, compared to their younger siblings – 77% of parents say their child has an influence always or most of the time (Table 4.13). Their influence grows, but more modestly, across the range of leisure pursuits; though perplexingly, their influence in choice of family restaurant stays at a modest 9%.

Teenagers' interest and influence in buying food is also double that of their younger siblings. Just under a quarter (23%) of parents say their child has an influence in what to eat most of the time. Certainly their knowledge of food grows as we have seen evidenced in the discussions above, so any suggestions or purchases they make are less 'risky'. Also, as parents will testify, food favourites are often an easy way to accommodate adolescents' moods and behaviour.

However, there are daily and practical reasons. As we saw earlier, adults' work patterns are changing, leading to a growing autonomy and responsibility among children. So-called 'latch-key kids' or single-parent teenagers, or parents' irregular work-patterns will leave children to prepare their own food. Children don't prepare food they dislike, so they

Table 4.13 Influence of 12–16-year-olds: how often does the child have any influence on the final decision made?

	% of parents saying influence always or most of the time
Kid's clothing	77
Choice of video film	36
Cinema film	34
Day out destination	26
Food	23
Holiday destination	21
Choice of TV programme	20
Family restaurant	9
Furniture item	4
Parent's clothes	2

Source: 'Kids as Customers' by James U. McNeal (USA); Henley Centre Planning for Social Change programme.

will more willingly play their role as chef where they have chosen their food. And this is likely to involve their favourite foods.

4.3.3 Children's favourite foods

A cynical observer might assume that a child's list of favourite foods would be a rank order of sugar content. However, children understand their health, and how food contributes to it. But favourites are favourites, and are indulgences, after all. Table 4.14 is revealing from four points of view.

- The list reveals a liking for the sweet and the sticky: chocolate is the most favourite, and ice cream/mousse fourth. The list shows a penchant for fried foods: chicken, chips, hamburgers, sausages and crisps make up half the top ten list. Favourites will include luxuries or extravagances, or the unnecessary: favourites, after all, are not what we eat all the time.
- The second relevation is a reasonable balance of otherwise healthy foods. Nutritionists would rightly take issue with the absence of some food types as they would with the inclusion of others: for example, green vegetables are not in the list (they are disliked by 39% of children), and nor is salad (disliked by 27% of children). While some foods may be fried, few would seriously quibble with the types of fried food, or with pizza or fruit or beef on the list.
- The third point is that the list of favourite foods includes four foods which children are not influential in buying. We should, after all, expect the list to contain only foods which children ceaselessly pester their parents to buy. Chicken is the second favourite, but is clearly more difficult for children themselves to cook – so their purchase influence is

Table **4.14** Children: their favourite foods (ranked in order of preference); and their purchase influence

	% purchase influence
Crisps	40
Sausages	15
Beef	10
Fruit	6
Hamburgers	35
Chips	35
Ice cream/mousse	31
Pizza	35
Chicken	10
Chocolate	33

Source: 'Kids as Customers' by James U. McNeal (USA); Leather-head Research; National Food Survey; FES; Asda; Henley Centre Planning for Social Change.

accordingly less. Perhaps fruit is ever-present in the home, and needs no special requests. Beef is not a convenience food, and is more likely to be a special occasion dish. Sausages are likely to be a regular quick family dish – with little need for special requests.

● The fourth revelation is that the list does, nevertheless, contain six foods where children have strong purchase influence – ranging from 31% of parents saying their child has strong influence in buying ice cream/ mousse, to an astonishing 40% influence in buying crisps. The consequence of such heady influence is that manufacturers have twin target markets in many food categories: adults and children. From our calculations, the direct influence of children on the spend in their top ten favourite foods is close to £2000 million a year.

4.3.4 Pocket money and income

Being a child has advantages, and one of them is that the influence of the child grows on what to eat, but someone else pays for it! This means that, as a child, you can keep your own money for what you really want!

The average weekly income for 5–16-year-olds – from pocket money and other earnings like newspaper rounds or family presents – is £3.86. The spending power of the country's youth adds up to some £34 million each week.

It is interesting to inspect how children spend their money. In the context of food, 41% of parents say their child spends money on crisps, sweets and ice cream, and we could further assume – where 29% of parents say the spend is on whatever the child likes – that some will be on food (see Table 4.15).

By and large, it is the responsibility of parents to look after nutrition. To that extent there is little change to the patterns of history. Where children are choosing and buying for themselves, there is naturally enough a pattern of self-indulgence: even savings are not so much against the rainy day, but

Table 4.15 Where does the pocket money go?: % parents saying their child spends money on . . .

	%
Crisps/sweets/ice cream	41
Savings	28
Comics/magazines	19
Records/tapes	15
Toys	13
Books/stationery	11
Clothes	9

Source: Walls Pocket Money Monitor.

for the more expensive items of their personal leisure (computer games, for example). Children develop early on a sense of what is family or public responsibility, and what is their personal pleasure.

4.3.5 Weight concerns

The list of ten favourite foods, while including the indulgent, also includes sensible foods. Children are aware of the relationship between the amount of food and weight gain, but sometimes to excess. It is perhaps surprising that 43% of 13–15-year-old girls are worried about their weight (Table 4.16). Even one-third (32%) of 11–12-year-olds, and a quarter of 9–10-year-old girls are worried. While boys up to 12 show less concern, by age 13–15 over a quarter (26%) are concerned.

Some girls do take their concerns too far, but for the majority of girls and boys it is a clear acknowledgement of the importance of food in health. We are not breeding premature neurotics, but rather a generation concerned to look good. Some commentators are concerned that girls' pursuit of role-model looks can be damaging to health, but the editorial stance of the magazines and fanzines is, by and large, sane and balanced, as is the eating advice given by the role-models themselves. At the end of the day, the sentiment of being worried about weight is a common enough expression, and seldom a genuine cry for help.

4.4 Model for change in children

The changes in each succeeding generation of children will be seen as a collision of random events, unless we can structure a model to explain the evolving circumstances.

Children of 30 years ago are similar in many ways to children of the late 1990s. Yet the differences are becoming more prominent as social and work structures change, and as the influence of the media pervades our lives, and recalibrates the role of family and parent.

Table 4.16 'I am worried about my weight'

Age (years)	Boys (%)	Girls (%)
5–6	8	4
7–8	7	20
9–10	7	25
11–12	16	32
13–15	26	43

Source: Asda 'Children and Food' survey (1993); Henley Centre.

- Parents are a source of knowledge, but increasingly not the only source.
- Parents are the stable role models for their children, but TV and sport build role models for children at every stage of their lives.
- Parents are the providers – of food, shelter, warmth, love. Increasingly food, shelter and warmth can be taken for granted.
- Parental love can never be replaced: nor the love that children need. But the affection that children can feel to bands or people is part of what the media offers as its very extended family.

The childhood of today's 9- or 15-year-old is markedly different to the childhood of his or her parents. The time children spend in leisure (i.e. time that is not taken up by school, sleep and other essential duties) is a circular four-stage process (Fig. 4.1).

4.4.1 Information and knowledge

Children have information and knowledge, and can be as informed as any adult. They share knowledge with parents about sport or food or the environment. They are players alongside people who play with them.

Thus, discovering and learning with other children at school is similar to other shared processes. With or without their parents, children watch football, a food programme, soaps, political satire, a war and the news – just in one evening. In children's eyes, are they not as qualified to comment on life as their parents? They have shared the same information and similar emotions.

In children's eyes, parents do not always qualify as better decision-makers: information changes; and children learn that our society is not feeding itself as well as it should. Kids themselves are more *au fait* with

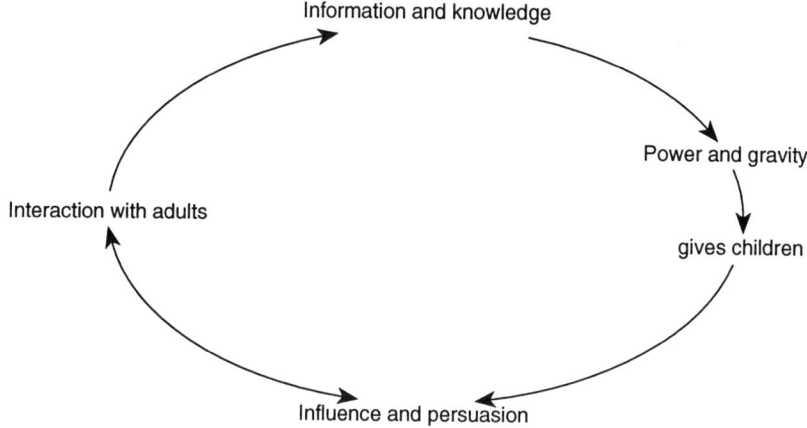

Figure 4.1 Children as people. (Source: Henley Centre).

kilojoules than their parents: all they lack is experience to frame their knowledge, and to give it a context.

4.4.2 Power and creativity

Information and knowledge bring for children a sense of control and power – just as they do among adults. Knowledge is power. And food is one area where children manipulate their power.

But it is not power and creativity in food alone: there are no kid-power revolutions in the kitchen! It is the attitude of mind, and the behaviour it brings. Children who share knowledge with their parents feel more powerful perhaps: but taking their own high ground fuels their sense of creativity. Power through the remote control; to form their own interactive world with a computer game (perhaps a mystery to parents); to zap between a score of TV channels; to cook their own food in the microwave; to e-mail and surf the Internet.

Knowledge is power, and technology is giving children the power and the creativity to share an adult world where it interests them. Where the adult world does not interest them, creativity adds dimensions to the world their parents knew as children. If, on the one hand, 'Cowboys and Indians' – a children's game – does not exist because people and horses limit the creative opportunities of technology and the imaginings of children, on the other hand, the fast-food restaurant down the road, or the microwave meal in the freezer puts children fair and square in the adult world.

4.4.3 Influence and persuasion

Once children are involved in the adult world, they buy the right to add their view. Every child knows that President Bush disliked broccoli: children have their world-famous anti-broccoli sponsor. Children of a certain age know about fibre and vitamins, what the body needs, that 'we are what we eat', why we need water, or what cholesterol is – and can leave their parents in befuddled ignorance.

Children, and more so teenagers, are therefore in a prime position to influence what the family eats and, certainly, what they themselves will eat. Their influence and persuasive skills are based on a wide knowledge and a range of people, from sporting heroes to club managers.

What is critical is the knowledge that children have, and the knowledge that they share with their parents. The effects of what basketballer X may say is diminished if a parent has no idea who he is: his physical and emotional prowess will carry little clout (even if he does like broccoli!).

So, the influence and persuasion skills of children conform to the adult pattern. It is easier to sell an adult a computer when he has a little knowledge (although he may not fully understand it) about RAM and

ROM and power and speed. It is equivalently easier for a child to persuade a parent, if between them they share some mutual appreciation. The media are influential: the two generations – unlike previous generations – meet on common ground, share the same emotions and interests, and are part of an audience that disregards age.

4.4.4 Interaction with adults

The fourth phase of the model 'children as people' is the interaction between children and adults. Shared knowledge and information are the key ingredients in this stage of the cycle. Children want to communicate; they want to interact. They have a built-in urge to learn and improve, because they want to be grown-up. Children may be no more wilful than adults, but are less diplomatic and less adept at exacting their needs.

So, interaction with adults and food? Children enjoy food, and have their own views; but equally the process is interactive. The more that parents show their interest, the more their children will emulate them. However, food, like sport, is for so many kids a *sine qua non* of living.

4.5 Conclusion

If adults had changed as much as children over the last three decades, we could argue that the development was a consequence of adding knowledge to experience in a continual process. That process does not exist with children. They are fresh into the world each day, and yet so quickly learn what their parents are learning and, at the same time, absorb it as an assumed part of life.

We have argued that children want to be part of the home. They are not on strike for independence. They are changing because of astonishing technological and media opportunities, and because their parents have work and living patterns that have evolved (as they will continue to).

Therefore, the fact that children are becoming ever more influential – our central theme – is not a trend to be explained away, but one to be accepted in the same way that we cannot imagine women not having the vote. Simply put, new patterns of life are developing new child behaviour and attitude patterns. This is clearly seen in the context of food. Child empowerment has made children's influence on food purchase and consumption prominent. The commercial importance of this will grow.

Reference

McNeal, J.U. (1900) *Kids as Customers.*

5 Children's nutrition: drivers for change

N. JARDINE and C. PHILPOTT

5.1 Introduction

Interest in the role of diet in the maintenance of good health and the prevention of disease is ever increasing. In this chapter we will examine how children's diets compare with the current dietary recommendations advised by many national authorities. Against this background we will examine some of the challenges facing food manufacturers who want to make products which help foster healthy lifestyles for children in the UK, through healthy eating habits.

Much of the emphasis in disease prevention is on reducing the levels of the 'diseases of affluence', the focus of the UK Government's *Health of the Nation* White Paper, published in 1992 (Department of Health, 1992). Many of these diseases strike older members of the population, but there is a growing belief that the origins of the diseases are seeded in poor nutrition experienced in childhood (BNF, 1991).

It is now axiomatic that the nutritional targets applied to adults are equally valid for children from the age of five years (Department of Health, 1992). However, children's eating habits differ from those of adults, evolving as they get older. To some extent the foods favoured by children have always differed from those which suit adult tastes – for example, children's dislike of 'greens' is notorious. But to improve the quality of nutrition, it will be necessary to change eating habits. This may not be any more straightforward for children than it is for adults, but the availability of suitable foods and food products available can only increase the chances of success.

Current dietary recommendations are exemplified by the population recommendations set by the panel convened to determine dietary reference values by the UK Government's Committee on the Medical Aspects of Food Policy (Department of Health, 1991). Besides the customary recommendations on levels of energy, protein, dietary fibre, vitamins and minerals, this report also suggested proportions of energy from carbohydrate and fat in the population's diet (Table 5.1). The dietary targets for fat and saturated fat in Table 5.1 have now been adopted by the UK Government as part of its Health of the Nation initiative.

Set against such targets, children's diets are often criticized. In

Table 5.1 COMA dietary recommendations

FAT	<35% of food energy
of which:	
saturated fatty acids	<11% of food energy
CARBOHYDRATE	>50% of food energy
of which:	
non-milk extrinsic sugars	<11% of food energy
intrinsic and milk sugars and starch	>39% of food energy

particular, criticism is levelled at many products, which appeal to children, for containing too much fat and sugar. There is also a belief that children do not consume enough vegetables and fruits, as a high consumption of these foods is believed to be protective against a variety of diseases. At the least, such concerns will maintain pressure on manufacturers to develop new products or modify existing ones to meet the perceived need for healthier food choice by children.

We will start with an examination of what we know about children's diets in the UK and their nutritional characteristics. We will then discuss the challenges involved in changing diets in the directions indicated on health grounds and some of the obstacles that will need to be overcome.

5.2 What are children eating today?

Numerous investigations have been made into the nature of British children's diets. However, the most useful studies are those which give a national perspective on children's diets in the UK. Two government reports have been published in the last seven years concerning children's diets. The first of these, a COMA report on the *Diets of British Schoolchildren* (Department of Health, 1989b) was published in 1989. This report, which we will call the 'schoolchildren report', dealt with school age children of 11/12 and 14/15 years. The second report, *The National Diet and Nutrition Survey: chldren aged 1.5–4.5 years* (Gregory *et al.*, 1995) was published in 1995 (the 'toddlers report'). Both contain a wealth of information on diets and their nutrient composition.

5.2.1 Energy

Children everywhere obtain their energy, macronutrients and micro-nutrients from a variety of sources but identifying these sources and comparing them between age groups presents difficulties since these two government surveys classify foods in different ways. Table 5.2 shows the major sources of energy listed in both surveys.

Table 5.2 Sources of energy in children's diets

Schoolchildren report	
bread	10%
chips	8%
milk	7%
biscuits	7%
other meat products	5%
(not carcass meats, offal, sausages, burgers or chicken in breadcrumbs)	
cake	5%
puddings	5%
other foods	52%
Toddlers report	
cereals and cereal products	*30%*
of which: breads	*9%*
breakfast cereals	*6%*
biscuits, buns, cakes and pastries	*9%*
milk and milk products	*20%*
of which: cows' milk	*15%*
vegetables, potatoes and savoury snacks	*12%*
of which: chips	*4%*
savoury snacks	*4%*
meat and meat products	*10%*
of which: sausages	*3%*
beverages	*8%*
of which: soft drinks (not low calorie)	*6%*
sugar, preserves and confectionery	*8%*
of which: chocolate confectionery	*6%*
fat spreads	*3%*
fruit and nuts	*3%*
fish and fish dishes	*2%*
miscellaneous	*2%*

Cereal products are prominent among the major contributors of energy. However, for most cereal products, the contribution of cereal products towards the energy intakes of British children is not a product of energy density. It is the large quantity consumed which determines the contribution towards energy intake.

5.2.2   Carbohydrates: starches and sugars

Carbohydrates are the major energy source in most adult diets and the same is true for children. Their potential importance to health, particularly of starchy carbohydrates, has only been widely appreciated in recent years. Unfortunately, the schoolchildren report does not list the major sources of carbohydrate, starches and sugars. However, the toddlers report goes into much more detail (Table 5.3). Half of carbohydrate comes from cereals and cereal products and beverages. Vegetables, potatoes and savoury snacks, milk and milk products and sugars, preserves and confectionery provide similar amounts, at around 12% each.

Table 5.3 Sources of carbohydrates in toddlers report

Total carbohydrate	
cereals and cereal products	*39%*
of which: breads	11%
breakfast cereals	9%
biscuits	6%
beverages	*14%*
of which: soft drinks, not low calorie	12%
vegetables, potatoes and savoury snacks	*13%*
of which: potatoes, fried	4%
savoury snacks	3%
vegetables, not potatoes	2%
other potatoes	2%
milk and milk products	*12%*
of which: cows' milk	9%
fromage frais and yoghurt	3%
sugars, preserves and confectionery	*11%*
of which: sugar confectionery	4%
chocolate confectionery	4%
sugar	2%
preserves	2%
fruit	*5%*
Starches	
cereals and cereal products	*64%*
of which: breads	11%
breakfast cereals	9%
biscuits	6%
vegetables, potatoes and savoury snacks	*26%*
of which: vegetables, not potatoes	3%
potatoes	3%
savoury snacks	1%
meat and meat products	*6%*
Sugars	
beverages	*25%*
of which: soft drinks	21%
fruit juice	2%
milk and milk products	*22%*
of which: cows' milk	16%
fromage frais and yoghurt	5%
cereals and cereal products	*19%*
of which: biscuits	5%
breakfast cereals	3%
breads	1%
sugars, preserves and confectionery	*17%*
of which: sugar confectionery	7%
chocolate confectionery	7%
sugar	3%
preserves	1%
fruit and nuts	*8%*

The main sources of starches are cereals and cereal products, which represent nearly two-thirds of starch intake. Vegetables and potatoes provide about a quarter of starch intake while a range of other foods provide smaller amounts.

Beverages and milk and milk products represent half of sugars intake. Cereals and cereal products and sugars, preserves and confectionery represent a third of sugars intake.

5.2.3   Fibre

Once again, cereal products are among the main contributors, providing nearly half of dietary fibre. They are closely followed by vegetables, potatoes and savoury snacks which provide over a third. Fruit and nuts contribute about 11% to fibre intakes. Table 5.4 shows the major contributors of fibre in toddlers' diets.

5.2.4   Fat

Table 5.5 shows the major sources of fat in children's diets. Milk naturally plays a far greater role in the diets of toddlers than of older schoolchildren. The older children obtained their fat from a wide range of sources, but the greatest single source remained milk. Meat and meat products were important contributors of fat in both toddlers' and schoolchildren's diets. Cereals and cereal products contributed a fifth of fat intake in the toddlers report.

5.2.5   Protein

For children in developed nations, obtaining an adequate supply of protein should not be a concern. Table 5.6 shows that protein is obtained from a wide range of sources for toddlers, with milk and milk products being a

Table 5.4 Sources of fibre in toddlers report

cereals and cereal products	*43%*
of which: breakfast cereals	14%
breads	13%
biscuits	5%
vegetables, potatoes and savoury snacks	*35%*
of which: vegetables, not potatoes	17%
potatoes, fried	8%
savoury snacks	6%
other potatoes	5%
fruit	*11%*
meat and meat products	*4%*

Table 5.5 Sources of fat in children's diets

Schoolchildrens report	
milk	11%
chips	10%
other meat products	8%
(not carcass meats, offal, sausages, burgers or chicken in breadcrumbs)	
biscuits	7%
carcass meats	7%
crisps	6%
butter	6%
other foods	47%
Toddlers report	
milk and milk products	*27%*
of which: cows' milk	21%
cheese	4%
cereals and cereal products	*20%*
of which: biscuits	7%
buns, cakes and pastries	4%
meat and meat products	*16%*
of which: sausages	4%
vegetables, potatoes and savoury snacks	*13%*
of which: savoury snacks	7%
potatoes, fried	5%
fat spreads	*9%*
chocolate confectionery	*6%*

Table 5.6 Sources of protein in toddlers report

milk and milk products	*33%*
of which: cows' milk	35%
cereals and cereal products	*23%*
of which: breads	8%
breakfast cereals	4%
meat and meat products	*22%*
of which: chicken and turkey	6%
beef, veal and dishes	5%
vegetables, potatoes and savoury snacks	*8%*
of which: vegetables, not potatoes	3%
potatoes	3%

particularly important source, providing about one-third of protein intakes. Cereals and meat products together provide nearly half of protein intake.

5.2.6 Vitamins

Taking two vitamins at random, vitamin C and riboflavin (a B vitamin), the major sources in schoolchildren's and toddlers' diets are shown in Table

Table 5.7 Sources of vitamin C and riboflavin in children's diets

Toddlers report: riboflavin	
milk and milk products	*51%*
of which: milk, whole, semi-skimmed and fat-free	41%
other milk and products	10%
cereals and cereal products	*24%*
of which: breakfast cereals	16%
meat and meat products	*8%*
sugars, preserves and confectionery	*4%*
Schoolchildren report: riboflavin	
(boys aged 11/12 years)	
milk	30%
breakfast cereals	21%
carcass meats	4%
other foods	45%
Toddlers report: vitamin C	
beverages	*50%*
of which: soft drinks	30%
fruit juice	20%
vegetables, potatoes and savoury snacks	*19%*
of which: vegetables, not potatoes	7%
fried potatoes	8%
other potatoes	4%
fruit and nuts	*15%*
milk and milk products	*8%*
Schoolchildren report: vitamin C	
(boys aged 11/12 years)	
potatoes, not chips	17%
chips	16%
vegetables	15%
fruit juice	15%
fruit	13%
other foods	24%

5.7. Unlike energy and the major nutrients, children tend to obtain their vitamins from a smaller range of food items. Three-quarters of 11–12-year-old boys' vitamin C comes from five sources. Breakfast cereal fortification plays a major role in supplying B vitamins.

5.2.7   Minerals

Table 5.8 shows the major sources of iron and calcium intakes in schoolchildren's and toddlers' diets. Fortified breakfast cereals make a major contribution towards iron intakes, as do bread and meat. Calcium is obtained from a limited range of sources, chiefly dairy and cereal products.

Table 5.8 Sources of iron and calcium in children's diets

Toddlers report: calcium	
milk and milk products	*64%*
of which: milk, whole, semi-skimmed and skimmed	51%
cheese	6%
yoghurt and fromage frais	6%
cereals and cereal products	*19%*
of which: breads	6%
breakfast cereals	3%
meat and meat products	*3%*
vegetables, potatoes and savoury snacks	*3%*
sugars, preserves and confectionery	*3%*
Schoolchildren report: calcium	
(girls 14/15 years)	
milk	30%
bread	14%
cheese	10%
puddings	5%
other foods	41%
Toddlers report: iron	
cereals and cereal products	*48%*
of which: breakfast cereals	*20%*
breads	*11%*
biscuits	*5%*
vegetables, potatoes and savoury snacks	*14%*
of which: vegetables	*7%*
potatoes	*5%*
savoury snacks	*2%*
meat and meat products	*14%*
milk and milk products	*6%*
Schoolchildren report: iron	
(girls 14/15 years)	
bread	13%
breakfast cereals	8%
chips	8%
other meat products	6%
(not carcass meats, offal, sausages, burgers or chicken in breadcrumbs)	
cascass meats	6%
biscuits	4%
other foods	55%

5.3   Health issues for children's diets

The food industry of the 1990s is faced with well-informed adult consumers raising knowledgeable children. Historically, dietary advice was concerned with dietary adequacy and the prevention of deficiency diseases. Much of the advice on offer today concentrates on preventing excesses. Eating to avoid the 'diseases of affluence' in the future requires consumers to think about nutrition and food choice in a different way. This message is without

Table 5.9 Percentage of energy from major nutrients in diets of children and adults

	McCance & Widdowson (Holland et al., 1991) (breastmilk)	Toddlers report			Schoolchildren report		Crawley (1993)	Diet and Nutritional Survey of British Adults (MAFF, 1994)	COMA Recommendations (DRVs)	
	Baby	Age 1.5–2.5	Age 2.5–3.5	Age 3.5–4.5	Age 10/11	Age 14/15	Age 16/17	Adults (incl. alcohol)	including alcohol	just food energy
CARBOHYDRATE	39.8	49.9	51.5	52	50.4	49.3	44.1	43	47	50
intrinsic and milk sugars and starches	39.8	32.6	32.2	32.3	—	—	30.6	25.8*	37	39
NME sugars	0	17.3	19.3	19.7	15[+]	—	13.5	16.2	10	11
PROTEIN	4.8	13.6	12.7	12.6	11.9	12.4	12.5	15.2	—	—
FAT	53.5	36.4	35.8	35.4	37.9	38.2	41.5	38.4	33	35
Saturated fat	23.5	16.9	16	15.5	—	—	—	16	10	11
FIBRE (as non-starch polysaccharide)(g/day)	0	5	6	7	—	—	13.9	—	18	18

*Starch + lactose
[+]Data from Rugg-Gunn et al. (1986) for 11–12-year-olds.

a doubt one of the most crucial and yet most difficult to communicate to adults, let alone to children. However, certain themes are enduring, such as moderating fat intake, increasing consumption of complex carbohydrates and fruit and vegetables.

Table 5.9 shows the contribution of the major nutrients to energy intake in relation to age. The COMA recommendations for the population are also shown for comparison. It is interesting to note the tendency for younger children to be closer to the COMA targets for fat than adults.

Despite this, there is a remarkable degree of consistency across the age range, but there are some marked contrasts to the recommended levels, especially for refined sugars and saturated fat. The public health focus is likely to remain on these nutrients and on starchy foods as well to some degree.

There has been a long-term decline in the intake of starchy foods like bread and potatoes which, to some extent, mirrors the decline in energy intakes. Nevertheless, starchy foods are still an important feature of the diet (Table 5.3), but the popular forms tend to come with other ingredients which are viewed more negatively, like fat (as in the case of pizzas, chips and crisps), or sugar or both.

5.4   The role of fat in children's diets

Fat provides more than twice as many calories as the same weight of protein or carbohydrate. In practice, this means that energy and fat are often closely related. In the UK over the past 50 years, absolute fat intakes have been declining (Department of Health, 1994). However, Fig. 5.1 shows that overall energy intakes have been falling slightly faster. The result has been a steady increase in the percentage of dietary energy

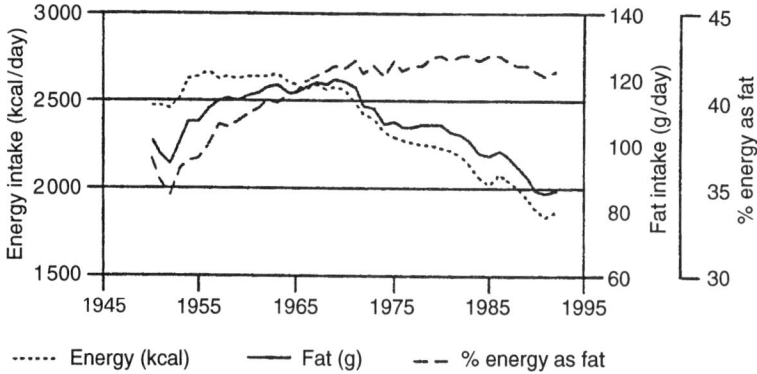

Figure 5.1 Fat and energy intake in the UK, 1950–92. (Department of Health, 1994).

derived from fat. There is some evidence that the trend has been arrested in the past two or three years and fat energy intakes are falling again, albeit slowly.

Today, average UK diets provide around 40% of energy from fat. This is too much, according to most dietary guidelines around the world. COMA advises that the population should be obtaining, on average, no more than 35% of food energy from fat and no more than 11% from saturated fatty acids (Table 5.1) (Department of Health, 1991). These targets have been incorporated into the Government's *Health of the Nation* White Paper (Department of Health, 1992).

A World Health Organisation study group, and certain governments around the world who have set population dietary targets, go even further and recommend that on average no more than 30% energy should be from fat with a maximum of 10% from saturated fatty acids (WHO, 1990).

This impetus to reduce fat comes from concerns about heart disease and overweight, two major health problems in developed nations (Department of Health, 1991, 1992; WHO, 1990). However, the relevance of these adult recommendations for children is often not clear.

For their size, children have high energy requirements. Figure 5.2 shows the high energy requirements per kg of bodyweight during early childhood and the subsequent fall as the child grows older. Adult energy requirements per kg of bodyweight are about half of that of an eight-year-old (Department of Health, 1991).

Children soon catch up with adults in absolute energy requirements. By the time they are eight, boys need as much energy each day as their mothers (Department of Health, 1991) and yet they are not of comparable

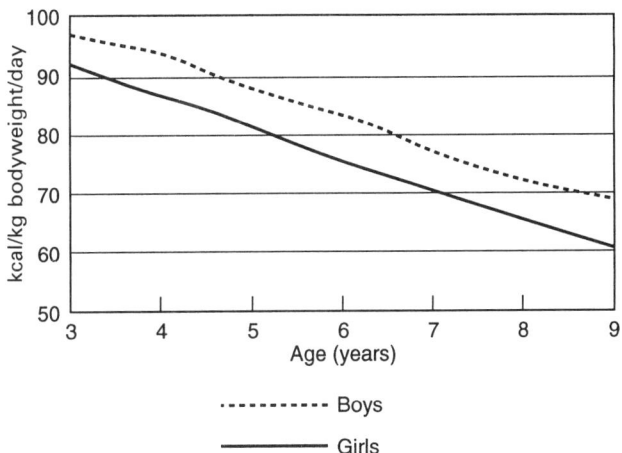

Figure 5.2 Children's energy requirements (kcal/kg bodyweight/day), 3–9 years. (Department of Health, 1991).

size. Children need to find ways of coping with the bulk of food that their energy requirements demand. One way is to eat frequently. This has other ramifications which are dealt with in the section on dental caries, but it also provides a way of getting large quantities of food into small children. In solving this problem of bulk, the close relationship between fat and energy can play a useful role, with high fat foods providing a concentrated source of energy.

Other sections of this book have discussed the requirements that children have from the food they eat. These requirements go far beyond physiological needs. Certain sensory aspects of foods particularly appeal to children, such as creaminess, melt-in-the-mouth sensations and aroma. All these characteristics rely, to a certain extent, upon the fat present.

The UK's COMA panel recommends that the guideline nutrient intakes for adults should be applied to children over the age of five years. However, this view is not shared by all. An advisory panel to the Canadian Government published a report (Health Canada, 1993) concerning children and dietary fat. In it they recommend that: 'From the age of two until the end of linear growth, there should be a transition from the high fat diet of infancy to a diet which includes no more than 30% energy fat and no more than 10% of energy as saturated fat'. Indeed, children begin life with a diet (in the form of breast milk) that provides over half of its energy from fat, so the only way is down. Table 5.9 shows that toddlers and schoolchildren have relatively lower fat intakes than adults. The key seems to be in preventing fat intakes from increasing during the teenage to adult years.

But this gradual reduction in fat intake cannot be successfully tackled by targeting children's favourite foods alone. A wide variety of foods contribute towards fat intakes, including those consumed at meal times (Table 5.5). Snack foods have a valid role to play in children's diets, providing pleasure and a range of useful nutrients.

5.4.1 Motivation to reduce fat intake

Perhaps the greatest incentive to reduce fat intakes is concern about weight. There is no doubt that obesity is a growing problem in developed countries (Dreon et al., 1988; Department of Health, 1994). Overweight is not socially or cosmetically desirable and overweight children may be teased at school. It is not surprising then that children are becoming more weight conscious. A preoccupation with weight can spill over into overzealous, even obsessional dieting. The incidence of eating disorders such as anorexia and bulimia in schoolchildren is on the increase, with girls as young as 9 and 10 reported to be dieting (Hill, Oliver and Rogers, 1992). Teenage girls are frequently found to have energy intakes below the recommended level. Low energy intakes imply low food intakes all round and such girls are often found to have low status in a range of other essential nutrients. Boys too are finding pressures to lose weight, even when they do not need to.

The food industry has a crucial role to play to project the image of food for pleasure and a healthy lifestyle. Associating products with healthy lifestyles seems to exist on two levels. Products such as isotonic drinks have a definite identity with sport and playing sport. However, this is a limited range of products aimed at a small segment of the population. Certain manufacturers of foods not directly linked with sport also manage to give exercise in general a very positive image, in turn strengthening their product image by associating it with a healthy lifestyle.

5.4.2 Reducing fat: the challenge to food manufacturers

To date, the food industry has met the challenge to reduce fat in a number of ways. One of these has been the introduction of reduced-fat alternatives to standard products, leaving consumers free to make their own choices. This approach has been particularly successful in the dairy sector. Nearly every dairy product now has a reduced alternative and consumers are accepting these new products. Reduced fat milk, for example, first became widely available in the late 1970s. Today, UK consumers now buy more reduced fat milk than whole milk (Fig. 5.3) (MAFF, 1994).

Products where technology has allowed reductions in fat include yoghurts, desserts, sausages, crisps, biscuits and ice cream. However for some product categories reducing fat still presents quite a challenge. Chocolate manufacturers, for example, are bound by legislation with rigidly defined compositional standards and are hampered by a lack of

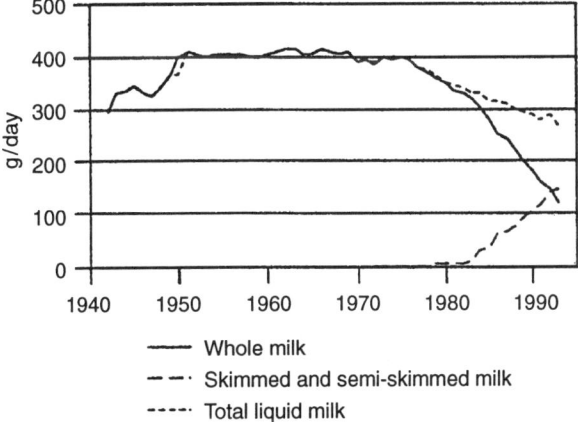

Figure 5.3 Consumption of milk in the UK, 1942–92. (Department of Health, 1994).

suitable alternatives for cocoa butter. Food manufacturers are also faced with indecisive legislation surrounding claims for reduced fat or 'light' products. A claims directive from the EU has been in the making for several years now with no early prospect of general issue.

Reducing saturated fatty acid intake presents a particular challenge to food manufacturers. High intakes of saturated fatty acids are linked with increases in blood cholesterol, a significant risk factor for heart disease. Despite the long-term nature of heart disease, it is thought that its origins may lie in childhood (Barker *et al.*, 1993).

The recommendation to reduce saturated fatty acid intake is equally applicable to children as to adults. However, with the types of food favoured in the UK, it is difficult to reduce saturated fatty acid intake without also reducing monounsaturated and polyunsaturated fatty acids in tandem. While this will contribute towards reducing total fat intake, these other fatty acids do have a useful role to play.

One way of approaching this dilemma would be to alter the balance of fatty acids. However, saturated fatty acids fulfil a number of crucial roles in products. They are mainly hard at room temperature, contributing towards solidification of products, and they are stable. Monounsaturated and polyunsaturated fatty acids are more commonly liquid at room temperature and would not fill the same role. Polyunsaturated fatty acids are also more prone to rancidity, thus reducing shelf life. So direct substitution of saturated fatty acids with monounsaturated or polyunsaturated fatty acids is often not a practical course.

5.5 Carbohydrates: starches and sugars

Carbohydrates are always in the news. The perceptions about them change, however, and this causes confusion. Traditionally, starchy foods have been regarded as the 'staff of life'. Perceptions became more negative a generation ago when starchy foods were seen as 'fattening', and advice was to take them in strict moderation. This is now known to be wrong and starchy foods have now come sharply back into fashion, at least in the nutrition community. A WHO study group suggested (WHO, 1990) that a population should obtain 50–70% of its energy from 'complex carbohydrates' derived from cereals, pulses and vegetables, and 40% has been suggested as a target for Scotland (The Scottish Diet, 1993). Current consumption levels are a long way short of these goals. In fact, children in the UK do better than adults, obtaining around 28% of energy from starch, whereas for adults it appears to be about 20% (Department of Health, 1991).

The simple carbohydrates, that is, sugars, raise even more controversy, at least when in a refined form. Over the years, a remarkable variety of ills has been ascribed to refined sugar, with the result that sugar has often been in headline news. Over the past decade, more sober scientific assessment (Glinsmann, Irausquin and Park, 1986; BNF, 1987; Department of Health, 1989a) has identified no major problems associated with sugar consumption at average levels of consumption, with the important exception of dental caries. Dental caries has declined considerably in recent decades even though sugar consumption has held up. This reduction is generally agreed to be due to the widespread use of fluoride especially in the form of toothpaste.

There is strong evidence that the frequency of ingestion is a much more important risk factor for caries than the amount taken in, although the two may be indirectly related. Despite this, and the decline in caries, even now nearly all dietary recommendations around the world suggest that it is the quantity of sugar consumed that should be reduced. A figure commonly chosen is that refined sugars (in the UK these are termed 'non-milk extrinsic' sugars) should contribute no more than 11% of food energy to the diet (WHO, 1990; Department of Health, 1991).

COMA recommended that carbohydrate should contribute 50% of dietary energy. In fact, children's diets in the UK are very close to this figure (Department of Health, 1989; Gregory *et al.*, 1995b). It is the balance of carbohydrates which deviates strongly from that recommended, so that actual intakes of starch and 'intrinsic' sugars are lower and refined sugars higher. It is interesting to speculate that this is partly because children start life having a natural tendency to like sweetness (Steiner, 1973). This is not surprising since 40% of the energy in breast milk comes from sugars, most of it in the form of lactose. While this sugar does not have the sweetness intensity of sucrose, its high concentration means that breast milk is sweet. Sweetness is therefore well accepted even before weaning.

5.5.1 Do current intakes of sugar by children matter?

There is no doubt that sugars are an important part of many foods which are well liked by children. Refined sugar attracts criticism on a number of grounds. The official view is that it is a major cause of caries and may also contribute to obesity (Department of Health, 1989a, 1994). There is also some concern that high sugar diets may be low in nutrient levels.

(a) Dental caries. The cause of dental caries is commonly ascribed to sugar consumption. However, the truth is more complex as can be seen by the fact that the incidence of caries has dramatically declined in developed countries, even though sugar consumption has remained at levels deemed

dentally unsafe. To understand the issues, we need to look at some of the factors involved in caries.

Bacteria in the mouth, especially those adherent to the teeth in the plaque, are able to break down sugars (that is, to ferment them) for energy and growth. Acids are produced as a by-product of this process and these can remove minerals from the enamel coating of teeth. This demineralization of enamel is to some extent reversible and saliva helps the process of remineralization, as does fluoride.

Three main factors contribute towards the cariogenicity of foods:

- carbohydrate content;
- frequency of consumption;
- oral clearance time.

Both the quantity and the type of carbohydrate in a food has a bearing on its cariogenicity. The main sugars used by oral bacteria to produce acid are those in foods and drinks, but starches are also broken down in the mouth to form sugars. Since many starches are retentive, they can give a steady supply of sugars to the oral bacteria. The amount of sugar required to give a high level of acidity at the tooth surface is not high – 10% sucrose in solution is sufficient to give the maximum depression in the pH (Imfeld, 1983).

Frequency of consumption is important in determining caries level in an individual. If the teeth are continually bombarded with carbohydrates and hence acid throughout the day, remineralization cannot occur and the result is a net loss of enamel. A related factor is oral clearance, that is the speed with which carbohydrates clear from the mouth. This will also influence the length of exposure of teeth to acid conditions.

It is not always easy to predict the cariogenicity of a food because of the number of factors involved. The acidogenicity alone is a poor guide. Foods can help protect teeth in various ways, for example by assisting the remineralization process, neutralizing acid or inhibiting the activity of oral bacteria. In addition, foods may stimulate saliva production, an important factor influencing dental health. The difficulties of predicting cariogenicity can be illustrated with two simple examples: bananas and raisins have a relatively higher cariogenicity than sucrose because they clear more slowly from the mouth. Secondly, milk chocolate has a lower cariogenicity than sucrose, possibly because certain compounds present in cocoa inhibit oral bacterial action (BCCCA, 1993).

From the foregoing, it will be apparent that caries is likely to be more affected by dietary patterns than by the actual foods consumed *per se*. Nearly all eating occasions include some form of fermentable carbohydrate. This could be starch from a breakfast cereal, bread with lunchtime sandwiches, sugar in a cup of tea or from a bunch of grapes. All these types of carbohydrate can be used by oral bacteria to produce acid.

Having said all this, the impact of diet on dental decay is overshadowed by the great decline in tooth decay which has been seen in developed countries over the last 20 years. This is almost certainly due to the effect of fluoride. In particular, the use of fluoride toothpastes has become almost universal, and this has been augmented in some areas by the fluoridation of water supplies.

The overriding importance of oral hygiene together with the use of fluoride toothpaste is now clear to health conscious individuals. Nevertheless, there is still a drive towards dentally safe food products particularly by dental educators. This has been given impetus by the setting up of standards for defining dentally safe food products by the Sympadent organization. However, the association confines itself to consideration of confectionery products. Founded in Switzerland in 1982, branches have since been set up in Germany and the UK.

The range of such products is dominated by sugar confectionery and even these are confined to a niche market at present. The reason is that, despite the advances in food technology in recent years, it is difficult to make reduced sugar products which match the sensory characteristics of the traditional products, particularly for picky consumers such as children. Chocolate products represents a particular challenge to confectionery technologists. The ingredients for reduced sugar products do not appear to be good enough yet for wholesale acceptance. The laxative side effects of many bulk sugar replacers – the sugar alcohols – may also be inhibiting sales.

(b) *Sugars and obesity.* There is a commonly held belief that sugar 'makes you fat', this criticism particularly being levelled at 'extrinsic' or refined sugars. In fact, there is no evidence at all to link sugar of any type with obesity. Indeed, studies with obese subjects have found that their diets are usually lower in sugars, and that they have a lower preference for sweetness than normal weight subjects (Drenowski *et al.*, 1992; Lewis *et al.*, 1992). A dietary survey from Scotland confirmed this finding. Researchers calculated nutrient intakes of a random population group. The group of subjects with the highest intake of fat had the highest incidence of obesity but this group also had the lowest refined sugars consumption (Bolton-Smith and Woodward, 1995).

These findings are not so surprising as they fit with a body of evidence that fatty foods are less effective in triggering satiety than foods rich in carbohydrate or protein and that it is more easy to overconsume fatty foods. From a metabolic standpoint, carbohydrate is not readily converted to fat in adipose tissue whereas absorbed fat can be directly laid down there. In addition, starchy foods tend to be bulky and this acts as an impediment to overconsumption. However, despite its obvious hedonic attractions, similar considerations appear to apply to sucrose. In a laboratory investigation, Green, Burley and Blundell (1994) compared the effect of snacks rich in fat or sucrose on satiety and energy intake. It was shown that energy intakes over a day were less when sucrose-containing snacks were consumed compared with fatty snacks but that effects on satiety were similar.

(c) Sugars and nutrient density.   One argument often used in favour of reducing current intakes of refined sugar is that sugar, containing energy but no micronutrients, dilutes the nutrients from the ingredients with which it is mixed. On the face of it, this appears to be a reasonable proposition, especially as it is possible to show that children who consumed high levels of sugar, when categorized according to sugar intake per unit of energy consumed, have lower levels of vitamins and minerals (Rugg-Gunn *et al.*, 1991). However, in practice, high sugar consumers also tend to have high energy intakes, so that overall intakes of micronutrients are not compromised. Indeed, high sugar consumers tend to have higher intakes of micronutrients (Rugg-Gunn *et al.*, 1991). This has also been shown in other studies for both children (Gibson, 1993) and adults (Black, 1991).

The COMA report on sugars summarized by noting that intakes of nutrients run parallel with total energy intake (Department of Health, 1989a). It also stated that: 'Sugars provide an acceptable means of contributing to the very high energy needs of some individuals.' As the report emphasizes, the main problem which could arise is where overall energy intake is low and sugar represents a high proportion of this energy, indicating the special importance of appropriate food selection in these circumstances. Aside from poor appetite due to illness, one of the most likely scenarios for a low energy intake is in teenage girls who have taken to dieting. This does represent something of a problem and one that is recognized, although food choice is ultimately in the hands of the individual.

(d) Sugar–fat see-saw.   There is one area where a form of nutrient dilution by sugar is very apparent, and one that is potentially helpful. This refers to the observation made repeatedly in dietary surveys carried out in developed countries that there is an inverse relationship between sugar and fat intakes (Botton-Smith and Woodward, 1994). For example, in a study in Cambridge, UK (Department of Health, 1989a), men and women were divided into tertiles according to consumption of refined (so-called 'extrinsic' sugars. The percentage energy from fat was calculated for each of these groups. Figure 5.4 shows that as the percentage of energy derived from sugars increased, so the proportion of energy from fats decreased.

As the UK COMA report on sugars acknowledged (Department of Health, 1989a), this is a general finding. It has also been widely observed in children (Rugg-Gunn *et al.*, 1991; Nicklas *et al.*, 1992; Gibson, 1993). The practical message is that people, whether adults or children, generally do not compensate for lower energy in fat or sugars by increasing their

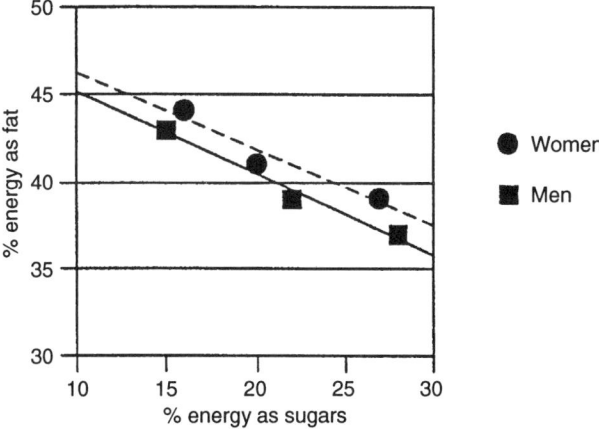

Figure 5.4 The sugar–fat see-saw (Department of Health, 1989a).

consumption of starch or for that matter protein. There is apparently no well-researched explanation why this is the case. A part of the explanation may be that starchy foods are bland and bulky. In quite different ways, fats and sugars, or perhaps sweetness, have particular attractions to the human palate. It is may be no accident that given unconstrained choice, human diets tend to converge to a level of energy provided by fat in the range of 35 to 45%. Perhaps those diets where fat is lower need some compensation as provided by sugars, be it on an organoleptic basis or related to energy density.

5.5.2   *Reducing sugar intakes*

Reducing sugar intake has frequently been a stated aim of health educators for a long time now. Such campaigns have had only limited success to judge from the contribution sugars make to diets of children and adults alike. As noted earlier, children begin life on a diet rich in a sugar – lactose – which provides nearly 40% of energy. By the age of around two years, the contributions of sugars to energy intake has reduced to around 30%. This is made up of 17% from refined sugars and 13% from intrinsic and milk sugars. However, if recommendations are to be followed, children are expected to make a transition to a diet containing 11% of energy from refined sugars. The question is, how realistic is this?

   To achieve this, the energy would have to be replaced by either fat, other carbohydrate in the form of starch or intrinsic sugars, or protein. Protein levels are already at recommended levels, and in any case protein intakes appear to be self-regulating as they are fairly constant at around

12–15% of energy intake from diet to diet or from age to age. Recommendations preclude increasing fat intakes as they are already higher than recommended. This leaves starch and intrinsic sugar intakes as the only ones which can be increased, and this is reflected in the COMA recommendations. However, Table 5.9 shows that starch and intrinsic sugar intakes are generally much lower than the recommended levels across all the age groups. The practical possibility of increasing these rests on children consuming more of starchy foods like bread and potatoes, and fruit. This would appear to be quite difficult, as many of the current starchy favourites bring with them either fat or refined sugar, for example, bread and butter, fried potatoes and some breakfast cereals.

5.6   Fibre

The role of fibre in the diet has been clarified during the past 20 years. One of the most important properties of fibre is its ability to absorb water, making the passage of food through the gastrointestinal tract smoother and faster. Its effectiveness in preventing digestive disorders such as constipation are well documented. But fibre also has the ability to bind up nutrients, preventing their absorption. This could prove to be a benefit if it binds up potentially harmful heavy metals but is detrimental if it binds up essential minerals.

There are no specific recommendations in the UK regarding fibre intakes for children. COMA recommends that children's fibre intakes should be proportional to adult intakes, based on their smaller size.

As a challenge to the food industry, encouraging healthy fibre intakes seems to be an easier product development proposition. Adding something to food is technologically simpler than taking something away. However, the problems associated with very high fibre consumption mean that the food industry cannot encourage excessive intakes through extensive fortification. Product development may not be the answer, though. Cereals and cereal products and vegetables, potatoes and savoury snacks currently make the greatest contribution towards fibre consumption, accounting for three-quarters of intake. Of cereal products, high fibre breakfast cereals represent a significant proportion of fibre intake, indicating that the nutritional message about fibre is being registered and acted upon.

5.7   Eating patterns and the nutritional role of snack foods

Children tend to eat more frequently than adults. The tendency is likely to relate to the activity levels of children, and their large energy needs in relation to body size. While many children will have set meals at breakfast, lunch and tea, they nevertheless have snacks of various types in between. These snacks often attract criticism on the basis that they tend to be rich in sugar or fat. Some favourite snack foods certainly do fall into these categories, but is snack consumption all bad?

It is unfortunate that the translation of nutritional guidelines into dietary advice often goes one step too far. Advice on a healthy diet becomes advice on the relative merits of specific foods. High fat, high sugar foods are derided, a perception reinforced by the media and health pressure groups, while children are commonly portrayed as consuming 'junk' diets, low in nutrients. Snack foods are commonly slated as being high in fat and/ or sugar and low in other nutrients such as vitamins and minerals. The nutritional value of children's preferred foods, such as chips, is often dismissed and yet the schoolchildren report (Department of Health, 1989b) found that chips contribute 20% to children's vitamin C and yet only 10% to fat energy intakes.

Some research has been done to examine the nutritional role of children's snacks. In a study of the snack eating habits of adolescents (12- and 15-year-olds), Robson and Strain (1991) showed that just under a third of energy came from snacks. The range of foods eaten between meals was quite extensive. The contribution to fat intake mirrored the energy intakes, but such snacks certainly tended to be high in sugars, contributing about 45% to consumption of all sugars (refined and intrinsic). Snack foods were found to contribute substantial proportions of iron (26%), calcium (32%), vitamin C (27%), thiamin (23%) and riboflavin (26%).

A more recent study (Summerbell et al., 1995) focuses more on the food commodities found in snacks. Snack energy was largely derived from confectionery, meat products, milk, biscuits, cakes, puddings and bread. This confirms the popularity of snacks containing sugars. Although this study does not analyse the implications of snacks for micronutrient intake, it confirms the diverse nature of foods used as snacks at all ages.

It also shows that as a group, snack foods contain proportionately less fat than the three main meals of the day (38% energy from fat for snacks and 41% for meals).

This study would suggest that there will be continued criticism of sugars in snack foods, if only on the basis of their contribution to the overall cariogenicity of the diet. However, it is also clear that probably the majority of children do snack between meals, and yet many have low caries experience. This highlights the protective role played by fluoride especially in toothpaste, and underlines that we have no reliable information about either the dietary habits or the oral hygiene practices of those children susceptible to decay. We therefore do not know where the balance lies between diet and oral hygiene. Clearly, ad libitum intake of potentially cariogenic snacks is to be discouraged on dental grounds, but it would be difficult to stop snacking entirely as it is now part of today's lifestyle, and such a pattern of eating follows children's inclinations to satisfy hunger fairly frequently. This area is likely to remain as a source of friction between manufacturers of foods used as snacks and health promoters.

5.8   Vegetable and fruit consumption

Another facet of health campaigns is a much increased emphasis on vegetable and fruit consumption. This is one area where the message has been consistent over many years, with an acknowledgement that intakes are lower in many developed countries than desirable. The importance of vegetables and fruit for health has been demonstrated in epidemiological work showing lower rates of cancer and heart disease in those who consume more of these items (Trichopoulos *et al.*, 1991; Block, Patterson and Subar, 1992; BNF, 1993). Parallel work in recent years has given a further boost to the importance of antioxidants, not only those vitamins found in plant foods but also the likely importance of other less well known substances with antioxidant properties.

As this research area develops, it is quite possible that the message to eat more vegetables and fruit will gain much more prominence, at the expense of other messages like reducing fat or sugar. We are already seeing campaigns in various parts of the world, exhorting consumers to 'eat five a day'. It is at least a positive message for once, and one which, if heeded, promises to do much to reduce disease and morbidity. As with other healthy eating messages, the public does not always know how to go about meeting the challenge. There is likely to be a great opportunity for those manufacturers who can make products based on plant foods which are convenient and attractive to eat. On average, intakes of fruits and vegetables are low so there is plenty of scope for new products which capture the imagination and the taste buds. One possible barrier which will have to be overcome is the perception that fresh is best. This is unlikely to be the case as it is clear that some preservation techniques like freezing and freeze drying immediately post-harvest can preserve the vitamins to higher levels than are found in less than fresh produce which has been stored for several days (BNF, 1987).

5.9   The outlook for the future and its product development implications

If current dietary recommendations are to be met by children, then at its simplest there will have to be a modest reduction in fat intakes, a larger reduction in refined sugar and a concomitant increase in starch intakes. Black (1991), in her analysis of the obstacles in the way of meeting dietary guidelines, comments that many of the foods rich in starches bring with them sugar and/or fat. The main foods available now which do not have this characteristic are bread, potatoes, rice and pasta. The amounts of these foods that would be required to make up for the loss of energy from reductions in fat and sugar in the diet are quite substantial, making a bulky diet likely to be daunting to today's children. The bulkiness of the diet is likely to be exacerbated if the recommendation to increase fruit and vegetables is also followed.

The UK population clearly has difficulty getting even near current guidelines as a recent analysis of data from *The Dietary and Nutritional Survey of British Adults* (Gregory *et al.*, 1990) shows. The figures should make chastening reading for those who are actively engaged in trying to change the nation's diet. Fourteen percent of people met the target for total fat but the target for saturated fatty acids was met by only 3%. Just 13% met the target for refined sugars. When the targets were combined, only 2% of men and 0.2% of women met the targets for sugars, total fat and saturated fat simultaneously. Even more striking is that only one man out of 1087 and none of the 1110 women met the combined targets for fat, saturated fat, sugar and dietary fibre (MAFF, 1994).

Clearly, adult eating habits have a very long way to go to meet current dietary guidelines. Against this sort of background, food manufacturers may well conclude that the eating habits of children are unlikely to change rapidly in the direction of dietary targets.

Putting recommendations and all arguments concerning their practicality aside, consumers are primarily interested in taste and children particularly so. One thing is certain, if new products designed to help children to meet dietary recommendations to not taste acceptable, then their success will be limited. It will remain a formidable challenge to food companies to make foods which are sufficiently attractive to eat that they become popular with children while at the same time helping them to meet dietary recommendations. Nevertheless, for those companies who take up the challenge, there should at least be the encouragement of public awareness brought about by public health campaigns like the Health of the Nation initiative in the UK.

References

Barker, D.J.P., Gluckman, P.D., Godfrey, K.M. *et al.* (1993), Fetal nutrition and cardiovascular disease in adult life. *Lancet*, **341**, 938–41.

Biscuit, Chocolate, Cake and Confectionery Alliance (1993), *A New Perspective on Diet and Dental Caries*. BCCCA, London.

Black, A.E. (1991), Sugars, fat and dietary counselling. In *Sugarless: The Way Forward*. Rugg-Gunn, A.J. (Ed.), Elsevier Applied Science, London and New York.

Block, G., Patterson, B. and Subar, A. (1992), Fruit, vegetables and cancer prevention: a review of the epidemiological evidence. *Nutr. Cancer*, **18**, 1–29.

Bolton-Smith, C. and Woodward, M. (1994), Dietary composition and fat to sugar ratios in relation to obesity. *Int. J. Ob.*, **18**, 820–8.

Bolton-Smith, C. and Woodward, M. (1995), Antioxidant vitamins adequacy in relation to consumption of sugars. *Eur. J. Clin. Nutr.*, **49**, 124–33.

British Nutrition Foundation (1987), *Food Processing: A Nutritional Perspective*, BNF, London.

British Nutrition Foundation (1991), Early Diet, Later Consequences. *BNF Bulletin*, **16**, Suppl. 1.

British Nutrition Foundation (1993) *Diet and heart disease: A round table of factors*. BNF, London.

Crawley, H.F. (1993), The energy, nutrient and food intakes of teenagers aged 16–17 years in Britain. *Br. J. Nutr.*, **70**, 15–26.

Department of Health (1989a), *Dietary Sugars and Human Disease*: Report of the Panel of Dietary Sugars, Committee on Medical Aspects of Food Policy. HMSO, London.

Department of Health (1989b), *The Diets of British Schoolchildren*: Report of the Sub-committee on Nutritional Surveillance, Committee on Medical Aspects of Food Policy. HMSO, London.

Department of Health (1991), COMA – Dietary reference values for food energy and nutrients for the United Kingdom. No. 41, HMSO, London.

Department of Health (1992), *Health of the Nation*. UK Government White Paper. HMSO, London.

Department of Health (1994), *Nutritional Aspects of Cardiovascular Disease*: Report of the Cardiovascular Review Group, Committee on Medical Aspects of Food Policy. HMSO, London.

Drenowski, A., Kurth, C., Holden-Wiltse, J. and Saari, J. (1992), Food preferences in human obesity: carbohydrates versus fats. *Appetite*, **18**, 207–21.

Dreon, D.M., Frey-Hewitt, B., Ellsworth, N. *et al.* (1988) Dietary fat: carbohydrate ratio and obesity in middle-aged men. *Am. J. Clin. Nutr.*, **47**, 995–1000.

Gibson, S.A. (1993), Consumption and sources of sugars in the diets of British schoolchildren: are high sugar diets nutritionally inferior? *J. Hum. Nut. Diet*, **6**, 355–71.

Glinsmann, W.H., Irausquin, H. and Park, Y.K. (1986), *Evaluation of health aspects of sugars contained in carbohydrate sweeteners*. Report from FDA's Sugars Task Force.

Green, S.M., Burley, V.J. and Blundell, J.E. (1994), Effect of fat- and sucrose-containing foods on the size of eating episodes and energy intake in lean males: potential for causing over consumption. *Eur. J. Clin. Nutr.*, **48**, 547–55.

Gregory, J.R., Collins, D.L., Davies, P.S.W. *et al.* (1995), *National Diet and Nutrition Survey: children aged 1.5 to 4.5 years* Volume 1: Report of the diet and nutrition survey. HMSO, London.

Gregory, J., Foster, K., Tyler, H. and Wiseman, M. (1990), *The Dietary and Nutritional Survey of British Adults*, OPCS Social Survey Division, HMSO, London.

Health Canada (1993), *Nutrition Recommendations Update: Dietary Fat and Children*. Joint Working Group of the Canadian Paediatric Society and Health Canada.

Hill, A.J., Oliver, S. and Rogers, P.J. (1992), Eating in the adult world: the rise of dieting in childhood and adolescence. *Br. J. Clin. Psych.*, **31**(1), 95–105.

Holland, B., Welch, A.A., Unwin, I.D., Buss, D.H., Paul, A.A. and Southgate, D.A.T. (1991), *McCance and Widdowson's The Composition of Foods*, 5th ed. The Royal Society of Chemistry and MAFF, London.

Imfeld, T.N. (1983), *Identification of low caries risk dietary components*. Karger, Basel.

Lewis, C.J., Park, Y.K., Behlen Dexter, P. and Yetley, E.A. (1992), Nutrient intakes and body weights of persons consuming high and moderate levels of added sugars. *J. Am. Diet. Assoc.*, **92**, 708–13.

Ministry of Agriculture, Fisheries and Food (1994) *National Food Survey 1993: Annual Report on Household Food Consumption and Expenditure*. HMSO, London.

Ministry of Agriculture, Fisheries and Food (1994), *The Dietary and Nutritional Survey of British Adults: Further Analysis*. HMSO, London.

Nicklas, T.A., Webber, L.S., Koschak, M. and Berenson, G.S. (1992), Nutrient adequacy of low fat intakes for children: the Bogalusa Heart Study. *Pediatrics*, **89**(2), 221–8.

Robson, P.J. and Strain, J.J. (1991), Snack energy and nutrient intakes of Northern Ireland adolescents. *Proc. Nut. Soc.*, **50**(3), 180A.

Rugg-Gunn, A.J., Hackett, A.F., Appleton, D.R. and Moynihan, P.J. (1986), The dietary intake of added and natural sugars in 405 English adolescents. *Br. J. Nutr.*, **51**, 347–56.

Rugg-Gunn, A.J., Hackett, A.F., Jenkins, G.N. and Appleton, D.R. (1991), Empty calories? Nutrient intake in relation to sugar intake in English adolescents. *J. Hum. Nut. Diet.*, **4**, 101–11.

The Scottish Home Office and Health Department (1993), *The Scottish Diet: Report of a Working Party to the Chief Medical Officer for Scotland*, The Scottish Office.

Steiner, J.E. (1973), *The gustofacial response: observations of normal and anencephalic newborn infants*. In *Fourth Symposium on oral sensation and perception: development in the fetus and infant*. Bosma, J.F. (Ed.). US Government Printing Office, Washington DC, pp. 254–78.

Summerbell, C.D., Moody, R.C., Shanks, J. *et al.* (1995), Sources of energy from meals versus snacks in 220 people in four age groups. *Eur. J. Clin. Nutr.*, **49**, 33–41.

Trichopoulos, D., Tzonou, A., Katsouyanni, K. and Trichopoulou, A. (1991), Diet and cancer: the role of case-control studies. *Ann. Nutr. Metab.*, **35S**, 89–92.

World Health Organisation (1990), *Diet, nutrition and the prevention of chronic diseases*. WHO, Geneva.

6 The politics of advertising to children
L. STANBROOK

6.1 Introduction

A key issue connected with advertising is whether advertising is harmful to children. Unfortunately, research is not a good moderator of the debate, since research cannot easily discover or qualify harm – especially harm shown to have been done over a period of several years. The concerns are specific: such advertising can be harmful because it might be responsible for a long-term negative behavioural and personality change in a child, or for damage done to interpersonal relationships, generally within the family, or, more controversially, as the unfair exploitation of young and gullible minds with inappropriate material. It is significant that this last concern brings the argument perilously close to notions of direct censorship.

In the past few years, consumer lobby criticisms of advertising to children have grown, predominantly in respect of the advertising of confectionery, snack food and soft drinks, but also to a lesser extent in respect of toy advertising. Worry has been expressed about the effects on children's diets and on family income as well as about the quantity and persuasive intent of such advertising.

Much of the criticism of children's advertising springs from worries about behaviour, in practice the key preoccupation of parents and guardians, who might not wish to see their children get too fat (or too thin), or to smoke, or to drink to excess, or to develop violent or anti-social characters, or to accept sexual stereotypes without question, or to become obsessive about other behavioural patterns or attitudes. The presentation of images or messages which either do not support or seem to contradict these preoccupations, therefore, become potentially subject to adult antagonism.

Such preoccupations can be encouraged by subjective interpretations of research results filtered through campaigning publications and the media. Where such interpretations seek to influence public policy, the very highest standards of examination and analysis should be brought to bear. This is, however, the very respect in which advertising to children has become a confused issue through politicization, resulting in such standards being either very low or non-existent. The politics – and the confusion – is evident even at the first stage of analysis. In Scandinavia, for example, the question of whether children's advertising should be restricted is not an issue. It is assumed that it should be restricted, because children's natural vulnerability makes such advertising inherently unfair (Goldstein, 1995).

While research is not generally able to examine and analyse the notion of harm, it is, however, able to show the rather different measure of advertising and other message comprehension among a specified audience.

Much of the research on children's advertising, in other words, is able to show that children understand and have an opinion of what is being communicated, but not whether the commercial communication is actually responsible for any behavioural changes other than in the strictly short term. Historically, research has been able to establish the existence of successful communication but would not seek to estimate its real behavioural effect. This is partly why such research has often emanated from within advertising agencies or within major companies. The commercial advertising interest requires in practice little more than the confirmation that the selected communications have been successfully made; that 'recall' is high among the selected audience. Advertisers are only too aware that the purchase decision itself is subject to several other factors which lie outside their influence.

In the children's confectionery market, surely among the fastest moving of all consumer goods, a critical factor is consumer reaction immediately following the initial consumption of the product. If a child does not like a particular product, no amount or quality of advertising will force a second purchase of the product in question. One would expect, nevertheless, the child to remember the brand name. Research relating specifically to the measurement of the successful communication of children's advertising has thus provided material for subsequent misuse as the apparent measurement of advertisements' long-term effect on children's behaviour.

As Jeffrey Goldstein has pointed out, all research is open to interpretation – and therefore open to question. As such it does not have unambiguous implications for public policy, despite its increasing application to this context (Goldstein, 1994). A tradition of research practice (planning) that started among researchers in order to measure advertising effectiveness has become corrupted by its misleading use within a different context altogether: that of public policy.

6.2 Research, public policy and politics

The requirements of public policy formulation, including health and social policy, rarely bring precision or clarity to the objectives of commercial communication. A political or campaigning approach designed to win over a particular political audience requires rather more than the measurement of recall levels or of opportunities to view. It requires research of a more fundamental character, in which advertising is invariably discovered to occupy a peripheral, rather than central, position, and in which assumptions about the overall process of education and socialization generally limit the effect of advertising to a barely significant level. Research of this variety is now increasingly commissioned. A good example of this newer type of research is *Influences on Children's Diet* undertaken by The Psychology Business Ltd (1994), using the technique known as attributional analysis.

The major findings of this research contradicted several established preconceptions about the influences on children's diet. It was found that families are much more focused on whether children will eat what is provided than on whether the proposed diet is nutritionally balanced. Furthermore, although families perceived TV advertising (correctly) as dominated by sweet snacks, they saw themselves as more powerful than the influence of advertising in making choices about food, and as having a high level of control in respect of nutrition and health (although concern about the former, significantly, was found to be at a low level).

The research estimated that TV advertising accounted for 7% of influences on children and 4% of influences on parents. Advertising was not seen as a distorting influence on the pattern of children's eating. Children did not continually pester their parents for specific advertised foods, and what children did ask for constituted a relatively small part of the behaviour that influences choice.

The political campaign waged against advertising is best analysed as a campaign against commercialism which simply uses advertising as the first and most obvious target. The political advocacy of the advertising interest, which primarily concentrates on the need to argue for the freedom of all commercial communications (about legal products and services), is called to account on behalf of the products and services it advertises. Since the range of these products and services is limited only by the fact of their legality, 'pure' advocates of advertising freedom are required to defend the freedom to communicate information about virtually everything that can legally be sold but are attacked because of assumptions about the products or services.

Such advocates are often critical of some of the products or services in question, but it is quite clear that the agendas of many public policy groups are assisted by the deliberate confusion in the public mind between the philosophical issue of advertising freedom as an aspect of free speech and the separate (and entirely political) issue of how to implement public (or private) policy objectives. Thus it is that those responsible for advertising are accused also of being responsible for tens of thousands of deaths by smoking, or for a variety of social and personal ills from lawlessness to obesity, alcoholism, dental caries, anorexia and coronary heart disease.

Thus it is also that the principle of free speech appears to be progressively subordinated to the requirements of consumer protection.

This level of debate owes much to public relations, a little to the abuse and corruption of research, and nothing at all to the rational discussion of public policy objectives. However, the parameters of the debate have become fixed by this depressing and corrupting interaction.

The substantial industry sectors thus threatened by advertising restrictions are beginning – only slowly – to understand that the 'game' is being contested by teams who are playing to very different rules. The 'game' is, of course, very serious indeed, since hundreds of thousands of jobs all over Britain and elsewhere in Europe depend upon the successful marketing of the products and services in these sectors both in domestic and export markets.

Research findings that show advertising to be an insignificant factor in, for example, dietary behaviour are dismissed immediately, sight unseen, when the research has been commissioned by industry or its representatives: 'They would say that, wouldn't they?' is a familiar cry. Independent research with similar findings, commissioned for example by Government departments, are also easily dismissed as a capitulation to industry influence. Very few of the public policy lobbies seem prepared either to examine, analyse or assess research from a properly scientific point of view or to accept legitimate criticisms of their own conclusions.

Instead, evidence that advertising is primarily responsible for anti-social or unhealthy behaviour is constructed from corrupt or discredited models, repackaged, re-pointed, re-interpreted to suit the political objective, and often even re-written. Consumer groups solemnly quote their own articles and monographs as statistical evidence to support their own political case. For example, Susan Dibb, the editor of *Children: Advertisers' Dream, Nutrition Nightmare* (1993), and adviser to the National Food Alliance, refers in this work to Surveys by the Food Commission (co-director, Susan Dibb) of a week's advertising during children's TV conducted in 1990 and 1992 and published in *The Food Magazine* (co-editor, Susan Dibb) under the heading 'A diet of junk food ads'. These publications are all adduced as research evidence to sustain amendments proposed by the National Food Alliance to the print and broadcast codes of practice in the UK. Thus, a series of mutually dependent half-truths and straight lies are resuscitated, repeddled and re-presented as accurate and scientific insights into the effects of advertising.

Some examples from the above-mentioned classic work in this genre: *Children: Advertisers' Dream, Nutrition Nightmare?* (Dibb, 1993), published by the National Food Alliance (NFA), an umbrella grouping of some consumer organizations, single issue lobbies, trade unions and societies in the food and advertising sectors.

The background to the 'options for more responsible advertising' is contained in the phrase: 'For the long term, (the Report) considers whether it is appropriate to target any advertising at young children.' This can be confirmed as the real political objective of the writers.

Because the 'they would say that, wouldn't they?' line has proved an insurmountable hazard for many independent corporate defence cases, individual companies are increasingly discovering the benefits of collective campaigning, and the industry's case justifying the continued recourse to children's advertising has largely been made through its trade associations.

A commentary on the NFA report was commissioned by the Advertising Association from Professor Patrick Barwise, of the London Business School. The commentary examined and analysed the references used in the NFA report, and concluded that it was 'dangerously one-sided and neglects most of the extensive published research on the issue of children and advertising, which is central to its argument . . . by exaggerating or misinterpreting much of the evidence and neglecting most of the rest, the report misrepresents the current and potential role of advertising in this area' (Barwise, 1994).

A further commentary was commissioned from Caroline Sharp (1994), an independent researcher. The commentary analysed every single reference made in the report and found that of those references which provided the main justification or basis of proposals for restriction, virtually none were correctly used or quoted. Many of the references were in fact untraceable.

Perhaps the greatest fault of the anti-advertising to children lobby in respect of food is that it attempts to view the world of children independent of the dietary trends and habits of the public at large. The focus of the Government's 'Health of the Nation' commitment, for example, is not children but adults. The main issue dealt with by the 'Health of the Nation', moreover, is not advertising but public health.

It is obvious from an examination of Government food consumption statistics that the dietary habits of the population at large have changed very significantly over the past 25 years, and that consumption patterns have indeed moved in the general direction favoured by Government dietary recommendations, without any necessity for proscriptive advertising controls (Advertising Association, 1993).

Over the past 25 years, despite a complete absence of Government controls over levels of advertising expenditure, consumers have spent less on products which Government reports have suggested should be consumed in lesser quantities. For example, consumer expenditure on red meat, many dairy products and white bread, has fallen sharply in real terms. In marked contrast, consumers have spent more on products recommended by Government agencies as 'healthier', such as fresh fruit, vegetables, poultry and brown bread. It is inconceivable that such large-scale changes in consumer attitudes and consumption habits would be confined only to adults and have left the children's market untouched.

The purpose of the anti-advertising case, in which *Children: Advertisers' Dream, Nutrition Nightmare* represented the benchmark, was to weave a huge patchwork of deliberate deceit; to formalize an array of dubious assumptions into apparently authoritative facts, the whole rendering virtually obscure the simple, unadorned and relatively unreported set of simple truths about what advertising can do and what it cannot do.

In the manufactured obscurity these truths keep fading from view in the struggle for public attention, and the mammoth deceit can be confirmed as the new dialectic. Where Government departments have been involved, this dialectic is quickly and disingenuously qualified as 'nationally agreed recommendations', qualifications which can often be enthusiastically but erroneously confirmed by over-zealous civil servants.

One of the key reasons explaining why these deceits have become successful is that the groups in question no longer subscribe to the ascetic and neutral principles of objective research. For them, research has ceased to be a trusted and independent arbiter and measurement of policy; it is rather an opportunity for exaggeration and hypostasis. It is the world of political PR, not that of research, in which the activist groups move and campaign. This world, in which industry interests in practice remain comparatively innocent – recognizes no independent standards save that of media coverage and lobby politics. Often the PR is discrete, intangible and subtle. It is also frequently dishonest.

6.3 Advertising: the front line

Advertising to children as an issue has now become part of the overall environment of public concern about children within society, bringing with it the emotionally charged public issues of education, security, parental responsibilities, health, youth culture and, increasingly, new media.

The issue of commercialism itself is under scrutiny, following years of determined deregulation in the UK. However generalized the real attack on the commercialized society may become, advertising and advertisements, especially in visual broadcast form, will almost always be the first targets because they are visible.

This visibility, particularly on TV, also renders them transparent. Advertisements normally fail when they are intangible, or too subtle or discrete. There is no marketing benefit whatsoever to be gained in confusing the potential customer. Unlike many political action campaigns, advertisements rely on transparent media and are paid for communications openly intended to inform and to persuade. Advertising's relationship with the independent media is inseparable – both display common characteristics within a free market environment. They are part of the general scramble – often unedifying, often undignified, never very subtle – for public attention. In such a scramble, it is hardly surprising that the elbows go flying and the politicians become jealous.

Advertising has always been an easy target. The introduction of commercial television in the UK coincided with well-documented public fears about conspiracies – from outside and from within. It fitted well with contemporary attitudes to suspect TV advertising of having a darkly hidden purpose, more sinister than door-to-door selling, quite possibly because interaction with the TV message was not available. It was a one-way communication; and therefore entirely different to the two-way discussions that could be had with the shopkeeper and the salesperson.

Advertising is in the front line because it provides the essential element in all marketing, namely communication. Albeit essential, it is nevertheless only one of the many stages that are necessary in getting a product or service to the consumer. Other stages are well-known: research and development, market research, design, packaging, production, distribution, retail and so on. Few politicians would actually contemplate restrictions on these stages in the way that advertising is so often positioned as the most appropriate element for restriction.

Advertising to children has become a primary focus of those who are concerned about the effects of advertising. This concern stems at least partly from the knowledge that advertising has an effect which competes in the scramble for the child's attention with other types of communications. Advertising's principal opposition, therefore, consists of all those groups, lobbies and causes whose own communications with the public, whether altruistic or not, appear to be impeded by commercial messages that may imply the opposite of the objectives of the groups concerned. It is irritating, for example, to a group concerned with the prevention of alcohol abuse to know that messages encouraging the consumption of particular alcoholic drinks compete for the attention of the general public.

However this irritation is expressed, various key aspects of this debate are often lost in a welter of emotion. These need to be better examined, particularly as most advertisers are generally happy to share many of the opinions and even assumptions of the lobby groups in relation to product concerns. A key premise of the advertising opposition is that advertising has a power and status that competes with parental guidance, educational influences or other interests not vested in the commercial equations of profit and loss. No research findings are ever offered for this premise, not only because they do not exist, but also because they are seldom, if ever, needed for the purposes of the campaign.

6.4   Gullibility

In the area of food advertising to children, the 'gullibility' angle is a crucial one. Advertisers are portrayed as cynical and selfish exploiters – as if in a dirty mac at the school gates – seeking to persuade children into practices which are bad for them. The frequent admonition of parents to children: 'Don't talk to strangers' appears to be undermined by the subtlety of the big advertiser who can insert messages into children's minds – and somehow ensure that those messages stick, as if advertising had succeeded in discovering the secret of brainwashing.

Gullibility – a presumed characteristic of innocence – is an emotive and essentially adult world, invariably used by patrons and parents. It is an experience that every adult can understand. This is perhaps because gullibility is something with which only adults identify: it is an adult word for a condition which only describes an adult mentality. Children rarely accuse each other of being gullible; the word they use is 'stupid' (or any of its many variations). It is also difficult, in practice, to assert that children can be gullible about advertising and yet also be firmly resistant to parental influence.

If consumer gullibility in commercial transactions is a function of innocence, then innocence is presumably the very last thing that anyone should encourage in children. 'Streetwise' children are likely to be good customers and consumers because they discriminate – because they are not innocent and not gullible.

6.5   Advertising as education

Advertising to children, before those children have the independent means to buy the goods or services advertised, is nothing less than primary education in commercial life; the provision, in effect, of free and elementary instruction in social economics – a passport to street wisdom. Far from being further restricted, as many suggest, this education course should in fact be supported, encouraged and enlarged.

Data concerning changes in advertising expenditure and in eating habits are fairly clear cut. However, the role of advertising in shaping and forming such changes is far less easy to show. Arguments can be advanced to suggest that the influence of advertising has been relatively small in relation to shaping the general trends of eating habits. The very high levels of television, newspaper and magazine coverage of dietary issues, together with Government recommendations and reports, have almost certainly been far more powerful influences on general dietary habits than food advertising, which is almost exclusively directed at the promotion of individual brands.

A major stated concern of the anti-advertising lobbies is that advertising messages may tend to undermine public health messages. This is based on a simple conviction: that the market has a political and social duty to implement government policies. It is related to a wider perception of the Government as the determinant of personal habits and lifestyles. This is of course a logical view – within a political environment of dirigisme. It would exclude a liberal and democratic state – and indeed this proposed exclusion has won the argument often in the past, for example in the Soviet Union and in Nazi Germany. It also exists – with difficulty – in the modern European political state. The French Government, for example, is currently seeking to persuade its EU partners to support a Declaration that the interests of health and consumer protection outweigh those of a single market in commercial communications. The British version first cropped up in the notorious comment of a Labour Minister in the late 1940s 'The men in Whitehall really *do* know best'.

The contention that advertising tends to outweigh other messages of a public service nature is false in most areas. All too frequently an analysis of advertising spend on a consumer product is compared with the equivalent advertising spend on public information, warning messages or other contra-indications in the presumed public interest about the use of these products. It is an entirely false measurement. There is a huge balance of editorial media coverage warning against the dangers of smoking, or which portrays smoking in a negative or disapproving way. This has a substantial collective effect which is all too often ignored. In practice, the public impact dwarfs any presumed market effect created by the isolated spots and messages of brand advertising.

It is now virtually impossible to believe that any smokers, even those who have recently started to smoke on a regular basis, are unalerted to the dangers of smoking. Many smokers are not only aware of the warning placed on packs and advertisements, but deliberately choose to ignore it or to disbelieve it. The right to take this view may be difficult to credit, but such credit has to be afforded. This is no doubt irritating, but it is a fact in a pluralistic and liberal society.

More recently, the explosion of media interest in dietary matters is having a similar effect on eating habits. In general, advertisers are relatively unconcerned about these editorial developments (except of course where the material is inaccurate or misleading): it is not part of their objective to seek to change people's dietary habits. In practice, therefore, there is no real competition between the presumed prescription and the contra-indication. Few people read or watch advertisements in the same way as they read or watch editorial or programming or examine food packaging. It is a false distinction in several ways, pointing to the possibility that all advertising which seeks to change socialized behaviour is wasted. Advertising intended only as education nearly always fails.

6.6   'Pester power' and parental responsibilities

Much advertising to children is of course nothing of the sort, since it is in fact directed to parents, even where it appears during children's programming. This is particularly true in respect of the more expensive toys, and it raises the spectre of what has been colourfully termed 'pester power'. While advertising to children is not supposed to encourage 'pester power' according to the restrictions in the British codes on TV and print media advertising, some connection with this concept is unarguable when the products advertised are either too expensive for pocket money or are food items to be consumed as part of the overall household food intake.

If 'pester power' is to be defined as including a child's request for an advertised product to be bought when the child, for various reasons, cannot make the purchase itself, then this is presumably what advertising to children should indeed be doing. At the same time, a considered parental response to the request is also exactly what parents should be doing. It is typically emotive to seek to qualify a child's request to parents or guardians for goods that have been advertised as 'pestering'. In practice, the phrase 'pester power' is frequently and wrongly ascribed to a scenario in which the advertiser 'pesters' the consumer.

But advertising lends the equivalent of a little finger to this influence, which is mostly derived from external influences as a whole, such as from reading and watching television (even the BBC), from school activities, from fellow schoolfriends, indulgent grandparents, guilty parents, from visits to the shops and so on.

Jeffrey Goldstein (1994) has analysed the difficulty that arises from the assumption, implicit in the Pester Power phrase, demonstrated by what he calls the Standard Argument, in which a TV commercial watched by a child creates desires which lead to demands, which lead to conflict between child and parent, which in turn can only be resolved by the purchase of the advertised item by the parent. His alternative approach postulates that peer influence creates desires. These desires lead to the selective viewing of commercials, which in turn leads to considered requests, at which stage a parental decision is made on the basis of these requests.

Recent criticism has seized on the extra advertising around Christmas, and the statistics showing this growth are triumphantly presented as conclusions in themselves. The issue of 'cumulative advertising' is presented as a potent source of public concern.

This is nothing more than empty thinking: a free rein given to the expression of irritation as a justification for statutory restriction. There is more advertising at Christmas because there are more sales at Christmas and because children receive presents at Christmas, for various sociological, psychological, and plain traditional reasons (Goldstein, 1994). No-one suggests that such advertising causes children's presents to be bought at this time of year. The advertiser's intent is to influence the choice of a planned purchase. The advertiser is thus competing with much more powerful sources of information, including the reams of feature material. In fact, it has been found repeatedly that advertised brands do not dominate lists of Christmas and birthday presents. Protesting about repetitive advertising at Christmas or submitting ingenious plans for autolimitation to the regulatory bodies is equivalent to protesting about raincoat advertising in the rainy season or about sun-tan oil advertising in the summer.

In the past, some children were denied access to books, dictionaries and encyclopaedias by their parents. This used to be a relatively common occurrence; some parents feared the effect of external influences on their children. Daughters, in particular, at an even earlier age, were often 'protected' from reading novels. Imagination in those days was an attribute deserving of the most profound parental suspicion. At the very extreme, some parents even tried to stop their children going to school because they feared that their innocent offspring would be corrupted by the outside world.

Our approach to this sort of behaviour should not change today just because the new demon appears to be commercial communication. In fact, quite the opposite. Media advertising in particular expresses itself under voluntary constraints which would be unimaginable in editorial copy. 'Pester Power' is a description of children's attributes and children's characteristics as seen exclusively by adults. It is not a description of a new persuasive force invented or generated by rapacious advertisers.

There is often common cause between the paternalism of consumer group agendas and the rather more justified paternalism of parents and teachers. Both wish to restrict the effects of commercialism. Parents can do this by being good parents. Teachers can do this by being good teachers. Consumer group activists can only do this by proposing statutory restrictions. These activists are essentially seeking to arrogate the parental role on behalf of officialdom. This represents a grotesque intellectual paradigm which is causing fundamental damage in the education process.

The first confectionery items purchased by children are generally the completely unadvertised 'penny chews'. As with toys, display is fundamentally important; a fact acknowledged by parallel consumerist campaigns to move confectionery displays away from checkouts.

Repetitive advertising is very irritating, especially if one concentrates, for example for research purposes, solely on the commercial breaks. But such advertising is not examined by the average child in the way that it is by the researcher. Several repeats of a particular advertisement are doubtless necessary to increase the likelihood of the advertisement being watched at least once by children whose concentration is known to be erratic.

Advertising to children will change significantly in the next decade, and the changes will be caused largely by new media. New ways will have to be found to reach the children's market, which some researchers claim has already gone underground as far as parents and guardians are concerned (Silvester, 1993). Popular culture in particular will always depend on successful advertising but may have to use different means for effective expression. Advertising structures will have to change. Consumer protection structures, equally, will also have to change.

The policy shift in focus from the critique of advertising in general towards that of advertising to children is perhaps unsurprising, following well-established trends in society towards the expression of greater public concern for children's welfare.

The policy shift is significant and revealing: advertising to adults has become less easy as a political target. The argument that advertising to the general public should be restricted for the good of the consumer has, in short, died for want of evidence. It is akin to the notion that politics and democracy should be limited to the educated classes. A satisfactory replacement option is to focus entirely on the protection of children on the basis of the gullibility issue. After all, children cannot vote. And apart from certain marginal forms of advertising, children are generally able to see or hear the same advertisements as adults. If children are to be hermetically sealed from all commercial communication, then of course no commercial communications should be allowed at all; except perhaps in pubs and betting shops (where, in general, they are already banned).

6.7   The Big Hijack: the role of government

Nowhere has this new focus on generalized proscription 'for the sake of the children' been better defined and elaborated than in the strange story of the attempted hijack of the Department of Health by a grouping of food policy and activist groups from early 1993.

There are two distinct pressures currently being exerted on those responsible for marketing to children. One is the erratic but frequently effective pressure of single-issue consumer group activity, evident also in a variety of other product areas. This activity consists largely of the utilization of variably-focused public sympathy and the manipulation of large areas of public and media ignorance about the role of marketing in order to make progress in precise, but often discrete, political objectives.

The other is the pressure exerted by the momentum of Government policy, characterized by innumerable working groups, standing committees, task forces, steering groups and other quasi-official groupings under the general heading of implementing or advancing Government policy requirements, largely but not exclusively in the area of public health policy.

The Government's concern about trends in public health was expressed in 1992 in its White Paper, the *Health of the Nation*, a document which set out various targets for improvement in the health of the English population. Similar documents or initiatives were prepared in respect of the rest of Great Britain.

The White Paper analysed various public health statistics and detailed a variety of targets to be achieved by concerted action over the next few years, in most cases by the year 2005. These included reductions in smoking, in obesity, and in the intake of fat as a percentage of food energy (from 40% to 35%) and of saturated fatty acids (from 17% to 11% of food energy).

Different elements of the Health of the Nation initiative were then passed to task forces, including the Nutrition Task Force.

While the issue of advertising, let alone advertising to children, received scant mention in the White Paper, the Nutrition Task Force gave itself a remit which included gathering and using the contributions of 'all those involved in a series of healthy food choices', and that media and advertisers should assist in 'giving the public information about diet, nutrition and health which encourages healthy eating'.

It was on this comparatively modest foundation that the Nutrition Task Force quickly decided (after rejecting requests from advertising trade associations to be represented in the group): 'to review current codes of advertising, particularly on TV with regard to children, in light of the HoN dietary targets, to consider how food advertising might help towards their achievement by reviewing available research findings and by initiating discussions with the advertising regulatory bodies'. It also set the objective to encourage industry to see healthy eating as a commercial opportunity by encouraging retailers 'to use advertising in a positive way, for example, through the promotion of healthy recipes and by liaising with the Nutrition Task Force Food Chain Working Group.'

These activities, worthy as they may be, will not improve the CHD rates in the UK within 5 or 10 years, which is what they are supposed to do, according to the original objectives of the White Paper and its associated comitology.

The relationship between trends in the nation's health and controls on advertising to children is not a particularly symbiotic one, but the connection with advertising interests has been established without further discussion as if it had been overwhelmingly obvious for years. In practice, it was a political hijack of an over-worked Government department, generating a false and scientifically useless connection between the health of the British adult population and the practice of TV advertising to children.

It is clear that the nation's health could be improved, and that children's diets may sometimes be unsatisfactory from a nutritional point of view. There are particular minority groups who may be at risk for a variety of different social and economic reasons, and who can be targeted in public information programmes, but TV advertising cannot solve these specific problems.

In fact, advertising already assists the progress of HoN objectives. There is a marginal and competitive benefit, both in advertising and product development, in promoting, producing and presenting 'healthy' food. No sociological developments in the next few years are likely to change this virtuous circle, just as no brand advertising is ever likely to enforce good dietary habits.

The two activities identified by the Nutrition Task Force (reviewing the codes and encouraging pro-health advertising messages) were to have provided the agenda for two working groups to be set up to identify ways of making progress. These have yet to meet, following protracted arguments between the National Food Alliance and the Department of Health over representation in the working groups. A letter from the National Food Alliance complaining that the groups were 'stuffed' with industry representatives was apparently all it took to kill any progress that might have been made in these areas.

Despite the lack of progress on the formal working groups, an informal working group of the Nutrition Task Force met in July 1994 to discuss the proposals on advertising in the Nutrition Task Force programme. For the first time, this grouping included representatives of the advertising industry. A further meeting was held in April 1996 which considered a review of the relevant research literature by researchers working on behalf of the Ministry of Agriculture. This review (Young, Webley, Hetherington and Zeedyk, 1996) was written to provide information on how TV commercials for food products affect children's choices (aged 8–12). It concluded that 'there is no evidence to suggest that advertising is the principal influence on children's eating behaviour.' Predictably, the National Food Alliance immediately discounted the review. Meanwhile, the Department of Health had quite clearly tired of the political machinations, and the Secretary of State wound up the Nutrition Task Force after its last meeting in October 1995.

The contention that industry should be encouraged to see healthy eating as a commercial opportunity is of course unarguable, but such encouragement is hardly necessary. Healthy eating has always been – and especially in the last two decades – a commercial opportunity for industry, which, together with retailers, have in many respects led the way in encouraging a better diet for the general public.

Only people can create a bad diet, by deciding to eat the wrong combinations and quantities of food on a regular basis. Even if people only ate food on the basis of seeking a balanced diet at all times, they could still eat snacks and confectionery, because these items can and do have a place in a balanced diet, as was confirmed by the detail of the *National Food Selection Guide*, published in July 1994.

It is wrong to assume that people eat exclusively for nutritional purposes; and it seems rather too zealous to demand that they should.

Fatty and sugary foods do indeed tend to be much more heavily promoted than fruit, vegetables or Colombian coffee. So are toys, especially at Christmas time. Such advertising of course comprises a larger proportion of all advertising directed towards children than of all advertising to all age groups. This is because children have simpler consumer interests and needs, and are unlikely for obvious reasons to be targets for the advertising of cars, financial services, expensive consumer durables and other products and services heavily advertised to older age groups. It is inevitable that lower cost items, such as snack food, cereals and dolls, will comprise a large proportion of advertising at times when children are the main audience.

A strategy for improved public health could best be addressed by a targeted programme of public education, encouraged by public authorities and supported by food manufacturers and retailers. Marketing restrictions, as proposed in some quarters, will not provide any contribution to the improvement of the nation's health. Indeed, it seems increasingly likely that the central problem – obesity – has nothing whatever to do with advertising. Noting that clinical obesity in Britain had doubled in the past decade, Prentice and Jebb (1995) showed that average recorded energy intake in Britain declined substantially over the same period. 'Evidence suggests that modern inactive lifestyles are at least as important as diet in the aetiology of obesity and possibly represent the dominant factor.'

The quality and characteristics of the debate have served to simplify the issues in some respects, especially in terms of public affairs considerations. The long-standing relationship between the Department of Health and the tobacco manufacturers, which produces on a regular basis an ever tighter set of advertising restrictions on tobacco, is an example of self-regulation by name only. It serves as a favoured model for the objectives of some Department of Health officials and consumer activists for the future control structure of food advertising to children.

The demise of the Nutrition Task Force has provided Government with an object lesson in public policy formulation, which might be summarised as: 'stick to the issues and don't get diverted.'

6.8 European political perspectives

There is no common policy in the European Union on the issue of advertising to children, although the sensitivity of the issue is acknowledged in most continental countries. Existing voluntary control systems and structures are inconsistent as between Member States, making it very difficult to show the same advertisement to children across the whole of Europe, or to abide by a properly and fairly administered system of common European advertising standards.

A recent study of the laws, regulations and control systems which bear directly on advertising to children in EU-12 showed that each member state has a different set of rules for the control of children's advertising (EASA, 1995). Public policy on the issue is essentially a matter of local cultural and political concerns rather than an objectively calculated means of protecting children. Much advertising restriction can be traced not to consumer or public concern but to simple domestic protectionism. The restrictions in Greece, for example, in which toy advertising may only be shown after 11 pm on television, wholly belong in this category, and the European Commission has recognized this by drafting legal proceedings against the Greek Government.

There is no common code of advertising practice which establishes common rules for the content of children's advertising, although the Toy Manufacturers of Europe have proposed such a common code to the European Commission, in its submission in respect of the Commission's Green Paper on commercial communications policy. This code is largely based on that elaborated in the United States, but has not been endorsed by most of those advertising food to children in the UK.

However, the European Union is reported to be close to drawing up a policy initiative in the area of protecting young consumers. Following widespread concern in Belgium and Italy over the selling of financial services to minors, a colloquium was held in November 1993 at the initiative of Melchior Wathelet, the Belgian Deputy Prime Minister, in the framework of the Belgian Presidency of the Council of Ministers, in co-operation with the Bureau Européen des Unions de Consommateurs (BEUC) and Belgian consumer and research organizations.

The colloquium was intended to provide benchmarks for policy development on the young European consumer using a conceptual framework (and colloquium sub-title) of 'Responsible Actor or Vulnerable Target?' (Bourgoignie, 1993).

There is scant European legislation which could come under the general heading of the consumer protection of young people. Directives on general product safety, toy safety and broadcasting appear to be the only examples of Union legislation in which the safety of children's commercial or consumer interests is addressed.

There are plenty of would-be legislators within the European Union institutions and agencies and the colloquium provided a perfect opportunity for the expression of the apparently obvious need to legislate at European level. The conclusions of the colloquium, scripted and announced with superhuman speed by Thierry Bourgoignie of Louvain University on the final day, confirmed the real aim of the colloquium. 'This conference' he said 'already points to a number of urgent matters of concern requiring intervention.' These included the control and supervision of 'unfair' advertising, which was defined, inter alia, as advertising to schools, joint offers, sales promotions and lotteries, and 'advertising establishing an artificial link between products and the social needs of young people' (Bourgoignie, 1993).

It was also proposed that the Commission should intervene in the area of food policy 'taking on board the consumer habits of young people'. This the Commission had already agreed: Christiane Scrivener, European Commissioner, had noted that the new Treaty Article 129a agreed as part of the Maastricht Treaty had installed consumer protection as a high priority in Community policy. She referred vaguely to new proposals to be prepared in 1994 for the young European consumer 'to create responsible and informed consumers.' These have not yet emerged from the Commission, but are known to be in active preparation.

Proposals for restrictions on advertising to children on television are currently being discussed in the European Parliament as it considers its approach to the Review of the Television Broadcasting Directive. The pressure for advertising restrictions comes largely from Sweden, which has only recently introduced commercial television and whose regulators are therefore deeply suspicious of commercial TV advertising. Restrictions on advertising to children need no justification in Sweden and Norway, and a complete TV ban on advertising to children is enforced in these countries.

Sweden in particular is anxious to avoid having to dismantle these controls and is therefore in the forefront of pressure to institutionalize its advertising ban throughout the European Union. In December 1995, it managed to persuade the Council of Ministers to agree a Declaration which appeared to sanction national broadcasting bans in respect of advertising. The real threat to the single market principle, however, remains within the possibility that the revised Broadcasting Directive might 're-nationalize' broadcasting control responsibilities back to the Member States from the European Union.

At the end of 1996, this prospect looks ever more likely. If the high point of European common policies in broadcasting has been reached (with the unwritten consensus among the EU Member States that more 'subsidiarity' should result in more control powers at national level for the restriction of television services), then the future economic health of the broadcasting medium in Europe looks to be in some doubt.

6.9 Advertising standards control

The two major regulatory bodies in the UK are the ITC and the ASA. The ITC is a statutory body, designed to regulate the broadcast medium. It includes representation from consumer bodies and the British Medical Association. The ASA is a self-regulatory body, set up by the industry, with independent membership, to administer the CAP Code and adjudicate on consumer complaints in non-broadcast media, including direct mail. The Code of Advertising Practice is designed by the industry working within the CAP Committee.

Both organizations have the power to levy severe sanctions, and the ASA has recourse to the OFT, where High Court injunctions can be obtained against persistent offenders. Both organizations use outside advisers and experts when appropriate.

There are extensive Independent Television Commission (ITC) conditions and restrictions on children's advertising (Section C.X of the British Code of Advertising Practice and Appendix 1 of the Code of Advertising Standards and Practice). Special protection for children is also provided in the Broadcasting Act. The ITC has recently agreed modifications of its rules on the advertising of food and slimming products, following a well-targeted and focused campaign by a coalition of consumer groups and health organizations. These modifications incorporated aspects not previously addressed explicitly: excessive consumption; the disparagement of good nutritional practice; and the targeting of children and adolescents in the advertising of slimming products.

As far as the printed media is concerned, new Codes of Advertising Practice and Sales Promotion were launched and implemented in January 1995. The new Code included a new rule specifically relating to food and children, which stipulated that advertisements 'should not actively encourage (children) to eat or drink at or near bedtime, to eat frequently throughout the day or to replace main meals with confectionery or snack foods' (see Appendix A).

The ASA also set up a working group to discuss the possibility of further modifications to its rules corresponding to those agreed by the ITC. However, the Codes, while similar, have never been exactly the same. Many of the changes agreed by the ITC were already included in the ASA code. The CAP Committee therefore rejected proposals for textual changes in respect of food advertising to children.

UK self-regulation works essentially on the basis of individual consumer complaints, and is not required or expected to be based on particular political or social agendas. The enforcement of such agendas would ultimately destroy the essentially democratic characteristics of the self-regulatory practice.

Regulation of advertising in the UK starts from a single and simple premise: that legally produced and sold goods should be allowed to be advertised freely subject to the substantial legal, regulatory and self-regulatory conditions in force.

Advertising on the broadcast medium is pre-vetted, at both script and film stage by the BACC (answerable to the ITC). The 'sensitive areas' of tobacco and health products are pre-vetted by the ASA in the print medium, but the remainder are not. (There are some 26 million advertisements published annually).

It is well-known that there are comparatively few complaints from the general public in the area of food advertising. A grand total of 32 complaints were made by the general public concerning nutrition claims in 1994; 11 were upheld. This is not, however, a cause for complacency. Codes are continually under review. It is inappropriate for consumer bodies directly to target the industry regulatory bodies as being responsible for delivering desired improvements in the nation's health; these bodies are not instruments of either government or lobbyist policy but rather are genuinely consumer based, and designed specifically to ensure that advertising remains legal, honest, decent and true.

There are also comparatively few consumer complaints about the advertising of food to children. Most of those that have been made have been presented by single issue consumer groups. Genuine individual complaints have been few and far between.

Self-regulation has qualities and attributes not always immediately understood by its critics. Among these are the following.

- Self-regulation, unlike statute, is cheap, fast and flexible. Codes can easily be rewritten or amended in response to changes in the barometer of public attitudes or assumptions.
- Self-regulation is honoured in the spirit as well as the letter. Statutory regulation has to cover all possibilities, not just those which are foreseen. As such it tends only to generate compliance with the letter of the law. The flexibility of self-regulation generates an environment, not just a code, of compliance with the rules. This works especially well in ensuring that advertisers are conscious of the rules even at the planning stage of advertising campaigns, and generates a reversal in the burden of proof. The system demands that advertisers supply evidence to support their claims.

The main codes of practice are all backed by law. There are some 80 separate pieces of UK legislation that can be used to enforce the codes, including the major legal instrument, the Misleading Advertisements Regulations 1988. Self-regulation is by no means a perfect mechanism that guarantees 100% compliance. This is one of the reasons why the statutory responsibilities given to the OFT in 1989 were welcomed by industry and consumers alike. But then neither does the imposition of statute law guarantee 100% compliance, whether in the UK or in other countries, where implementation and compliance are often battles to be fought almost as strongly as the passage of the legislation itself.

The central area of examination identified by the Nutrition Task Force would appear to be the degree to which the advertisers and the regulators of advertising are prepared to condition their practice in respect of the recommendations of the *Health of the Nation* White Paper and, by further implication, the programme of the Nutrition Task Force, as against the existing conditions imposed by law and self-regulation. It remains the case, however, that the Codes do not, and are not intended to provide the minimum requirements in children's advertising standards.

The greatest area of self-regulation is present at the initial stage of campaign planning, asserting itself as a function of the relationship between the client and the agency. Each knows and respects the unwritten rules of a free and responsible market. This generates powerful penalties for irresponsible advertising, namely, loss of trade and turnover, loss of accounts, distribution and licensing agreements. It is not in the advertiser's interest to engage in advertising or marketing which is misleading or false: such advertising is easily unmasked and exposed by an increasingly demanding, selective and sophisticated consumer.

Those who do not respect these rules will not be able to sell to children. This is not a first draft for a Government edict. This is a market signal.

Appendix A: The ITC Code of Advertising Standards and Practice:

Advertising and Children; Particular care should be taken over advertising that is likely to be seen by large numbers of children and advertisements in which children are to be employed.

The Child Audience	1	At times when large numbers of children are likely to be viewing, no product or service may be advertised, and no method of advertising may be used which might result in harm to them physically, mentally or morally, and no method of advertising may be employed which takes advantage of the natural credulity and sense of loyalty of children. For the purposes of this Code, unless otherwise stated, the Commission normally regards as children those aged 15 years and under.
Misleadingness	2	Children's ability to distinguish between fact and fantasy will vary according to their age and individual personality. With this in mind, no unreasonable expectation of performance of toys and games may be stimulated by, for example, the excessive use of imaginary backgrounds or special effects.
	3	Advertisements for toys, games and other products of interest to children must not mislead, taking into account the child's immaturity of judgement and experience. In particular:

(a) The true size of the product must be made easy to judge, preferably by showing it in relation to some common object by which it can be judged. In any demonstration it must be made clear whether the toy can move independently or only through manual operation.

(b) Treatments which reflect the toy or game seen in action through the child's eyes or in which real life counterparts of the toy are seen working must be used with due restraint. There must be no confusion as to the noise produced by the toy – eg a toy racing car and its real life counterpart.

(c) Where advertisements show results from a drawing, construction, craft or modelling toy or kit, the results shown must be reasonably attainable by the average child and ease of assembly must not be exaggerated.

Competitions   4   If there is to be a reference to a competition for children in an advertisement, the published rules must be submitted in advance to the licensee. The value of the prizes and the chances of winning one must not be exaggerated.

Direct Exhortation   5   Advertisements must not exhort children to purchase or to ask their parents or others to make enquiries of purchases.

Appeals to Loyalty   6   No advertisement may imply that unless children themselves buy or encourage other people to buy a product or service they will be failing in some duty or lacking in loyalty.

Inferiority   7   No advertisement may lead children to believe that if they do not have or use the product or service advertised they will be inferior in some way to other children or liable to be held in contempt or ridicule.

Direct Response   8   No advertisement may invite children to purchase products by mail or telephone.

Restriction on Times of Transmission   9

(a) Advertisements for the following must not be transmitted during children's programmes or in the advertisement breaks immediately before or after them – alcoholic drinks, liqueur chocolates, matches, medicines, vitamins or other dietary supplements, 15 and 18 rated film trailers.

(b) Except in circumstances approved by the Commission, the following will be acceptable only after 9 pm:
   (i) advertisements in which children are shown having any medicine, or vitamin or other dietary supplement administered to them;
   (ii) advertisements for medicines, or vitamins or other dietary supplements which use techniques that are likely to appeal particularly to children, such as cartoons, toys or characters of special interest to children.

(c) Children must not be shown self-administering medicines or vitamins or other dietary supplements unless prior permission is given by the Commission.

NOTES:

(i) *For the purposes of this rule 'medicines' are classified as products which carry a product licence. 'Dietary supplements' are classified as isolated or highly purified or concentrated products sold in forms resembling medicines, eg. vitamins, minerals and amino acids.*

    (ii)   *In the case of a product which cannot easily be distinguished from a medicine, or where the advertising itself contributes to such a lack of distinguishability, particularly with regard to very young children (those five years old and under), Rule 9(a), (b) and (c) above should be applied.*

    (iii) *Where an exemption is sought under this rule it is likely to be granted only in relation to products such as those for oral hygiene, skin preparations including acne treatments and externally-applied decongestants. The exemption will be granted only if the Commission is fully satisfied that harm is unlikely to arise as a result of very young children's responses to the advertisement.*

(d)   Advertisements in which personalities, or other characters (including puppets etc) who appear regularly in any children's television programme, present or positively endorse products or services of particular interest to children must not be transmitted before 9 pm. This does not apply to public service advertisements or to characters specially created for advertisements.

(e)   Advertisements for merchandise based on children's programmes must not be broadcast in any of the two hours preceding or succeeding transmission of the relevant programme or of episodes or editions or the relevant programme.

(f)   Advertisements which contain material which might frighten or cause distress to children must be subject to appropriate restrictions on times of transmission designed to minimise the risk that children in the relevant age group will see them. Trailers for 15 or 18 rated films must not be shown in or around children's programmes and, depending on content, may require more rigorous timing restrictions.

Prices   10   Except in the case of services carrying advertising directed exclusively at audiences outside the UK, advertisements for expensive toys, games and similar products must include an indication of their price.

(a)   A product will not be regarded as expensive if it is reasonably widely available at a retail price below that specified by the Commission from time to time.

(b)   Where a range of products is featured in a single advertisement only the most expensive item need be priced.

(c)   Where more than one item is priced, each price must clearly refer to a particular item.

(d)   When parts, accessories or batteries which a child might reasonably suppose to be part of a normal purchase are available only at extra cost, this must be made clear.

(e)   The cost must not be minimised by the use of words such as 'only' or 'just'.

Health and Hygiene   11

(a)   Advertisements must not encourage children to eat frequently throughout the day.

(b)   Advertisements must not encourage children to consume food or drink (especially sweet, sticky foods) near bedtime.

(c)   Advertisements for confectionery or snack foods must not suggest that such products may be substituted for balanced meals.

Safety   12   Any situations in which children are to be seen or heard in advertisements should be carefully considered from the point of view of safety and it should be borne in mind that, in some circumstances, bad examples set by adults may also encourage dangerous emulation.
In particular:

(a)   **Road Safety**
(i)   children must not appear to be unattended in street scenes unless they are obviously old enough to be responsible for their own safety;
(ii)   children must not be shown playing in the road;
(iii)   children must not be shown stepping carelessly off the pavement or crossing the road without due care;
(iv)   in crossing busy streets, children must be seen to use pedestrian crossings;
(v)   children must behave in accordance with the Highway Code, whether as pedestrians, cyclists or passengers.

NOTE:
Under the Motor Vehicles (Wearing of Rear Seat Belts by Children) Act 1988 and the Motor Vehicles (Wearing of Seat Belts by Children in Front Seats) Regulations 1993, children under 14 must normally wear seatbelts in vehicles which are fitted with them.

(b)   **General Safety**
(i)   children must not, for example, be seen leaning out of windows, climbing or tunnelling dangerously, or playing irresponsibly in or near water;
(ii)   small children must not be shown climbing up on high shelves or reaching up to take things from a table above their heads;
(iii)   medicines, disinfectants, antiseptics and caustic or poisonous substances must not be shown within reach of children without close parental supervision, nor may children be shown using these products;
(iv)   children must not be shown using matches or any gas, petrol, paraffin, mechanical or mainspowered appliance which could lead to them suffering burns, electrical burns, electrical shock or other injury;
(v)   children must not be shown driving or riding on agricultural machines (including tractor-drawn carts or implements); scenes of this kind could encourage contravention of the Agriculture (Safety, Health and Welfare Provision) Act 1956;
(vi)   an open fire in a domestic scene in an advertisement must always have a fire-guard clearly visible if a child is included in the scene.

Danger   13   No advertisements may encourage children to enter strange places or to converse with strangers (for example, in an effort to collect coupons, wrappers, labels, etc). The details of any collecting scheme must be submitted to the licensee who must be satisfied that it contains no element of danger to children.

Exploitative   14   Advertisements must not portray children in a sexually **provocative** manner. Treatments in which children appear naked or in a state of partial undress require particular care and discretion.

Clubs	15	No advertisement dealing with the activities of a club may be accepted without the submission of satisfactory evidence to the licensee that the club is properly and responsibly supervised.
Good Manners and Behaviour	16	Children in advertisements should be reasonably well-mannered and well behaved.
Children as Presenters	17	Children must not be used formally to present products or services which they could not be expected to buy themselves. This applies whether or not such products are of interest to them. Nor may they make in relation to any product or service, significant comments on characteristics of which they cannot be expected to have direct knowledge.
Testimonials	18	Children must not be used to give formalised personal testimony. This does not, however, preclude children giving spontaneous comments on matters in which they would have an obvious natural interest.

Appendix B: Advertising Standards Authority; The British Codes of Advertising and Sales Promotion (excerpt)

Children

1 The way in which children perceive and react to advertisements is influenced by their age, experience and the context which the message is delivered. The ASA will take these factors into account when assessing advertisements.

2 Advertisements and promotions addressed to or featuring children should contain nothing that is likely to result in their physical, mental or moral harm:

a they should not be encouraged to enter strange places or talk to strangers. Care is needed when they are asked to make collections, enter schemes or gather labels, wrappers, coupons and the like

b they should not be shown in hazardous situations or behaving dangerously in the home or outside except to promote safety. Children should not be shown unattended in street scenes unless they are old enough to take responsibility for their own safety. Pedestrians and cyclists should be seen to observe the Highway Code.

c they should not be shown using or in close proximity to dangerous substances or equipment without direct adult supervision. Examples include matches, petrol, certain medicines and household substances as well as certain electrical appliances and machinery, including agricultural equipment

d they should not be encouraged to copy any practice that might be unsafe for a child.

3 Advertisements and promotions addressed to or featuring children should not exploit credulity, loyalty, vulnerability or lack of experience:

a they should not be made to feel inferior or unpopular for not buying the advertised product

b they should not be made to feel that they are lacking in courage, duty or loyalty if they do not buy or do not encourage others to buy a particular product

c it should be made easy for them to judge the size, characteristics and performance of any product advertised and to distinguish between real-life situations and fantasy

d parental permission should be obtained before they are committed to purchasing complex and costly goods and services.

4 Advertisements and promotions addressed to children:

a should not actively encourage them to make a nuisance of themselves to parents or others

b should not make a direct appeal to purchase unless the product is one that would be likely to interest children and that they could reasonably afford. Mail order advertisers should take care when using youth media not to promote products that are unsuitable for children

c should not exaggerate what is attainable by an ordinary child using the product being advertised or promoted

d should not actively encourage them to eat or drink at or near bedtime, to eat frequently throughout the day or to replace main meals with confectionery or snack foods

e should not exploit their susceptibility to charitable appeals and should explain the extent to which their participation will help in any charity-linked promotions.

5 Promotions addressed to children:

a should not encourage excessive purchases in order to participate

b should make clear that parental permission is required if prizes and incentives might cause conflict between children and their parents. Examples include animals, bicycles, tickets for outings, concerts and holidays

c should clearly explain the number and type of any additional proofs of purchase needed to participate

d should contain a prominent closing date

e should not exaggerate the value of prizes or the chances of winning them.

References

Advertising Association (1993), *Review of Arguments and Data relating to the NFA Report*, Advertising Association, p. 9.

Barwise, P. (1994), *Children, Advertising and Nutrition*. London Business School and Advertising Association.

Bourgoignie, T. (1993), *The Young European Consumer: Responsible actor or vulnerable target?*, Academia, Brussels, pp. 55–62.

Dibb, S. (Ed.) (1993), *Children: Advertisers' Dream, Nutrition Nightmare?*, National Food Alliance.

EASA (1995), *Draft Alliance Report on Self-Regulation for advertising & children in Europe.* European Advertising Standards Alliance, Brussels.

Goldstein, J. (1994), *Children and Advertising: Policy implications of scholarly research*, The Advertising Association.

Goldstein, J. (1995), *Children and Advertising in Scandinavia*, prepared for the Toy Manufacturers of Europe.

Prentice, A.M. and Jebb, S.A. (1995), Obesity in Britain: Gluttony or Sloth? *B.M.J.*, **311**, 437–9.

Sharpe, C. (1994), *An analysis of the references used in Children: Advertiser's Dream, Nutrition Nightmare?*

Silvester, S. (1993), 'Eurokids'. In *The Young European Consumer: Responsible actor or vulnerable target?* Academia, Brussels, pp. 25–37.

Young, B., Webley, P., Hetherington, N. and Zeedyk, S. (1996), *The role of television advertising in children's food choice*, Ministry of Agriculture, Fisheries and Food.

7 How packaging works with children

S. H. L. CLARK

7.1 Introduction

It is rare to find an objective evaluation of the impact of packaging on the perception and behaviour of any target market, children or otherwise. Many companies rely on subjective judgement – located somewhere between the brain and the gut, with a detour through the heart.

However, I can think of three objective methods of evaluating how packaging works with children:

1. we can watch and observe their behaviour in the retail environment;
2. we can ask them questions and listen to the answers;
3. we can change packaging and measure the result.

I shall try and avoid subjective judgements and confine my comments to objective measurements, with liberal doses of examples, which I hope will illuminate – if not prove – how packaging works with children.

Before moving on to examples, let me explain the theory of shopping 'scripts', and the role of packaging in the retail environment.

7.2 The role of packaging

Packaging plays a vital part in the brand's overall communications, particularly as the brand's main, often only, spokesperson within the crowded shopping environment.

To maximize its effectiveness, it must exploit the entire pack format using shape, structure, materials as well as name, graphic elements, and pack copy.

The criteria for developing the optimum packaging design must be based on clear communication of relevant brand values together with appropriate codes and triggers that will aid the purchase decision. Both of these must be developed with a view to what best fits the consumer's particular shopping mode within a specific market. 'Script Theory' can help identify both the relevant shopping mode and its implications for pack design.

7.2.1 *Script theory and brand choice*

In order to minimize the confusion and chaos that would be caused by the need to make new decisions every time we undertake a task or activity, we develop unwritten 'scripts' to help define and guide our behaviour. In using these 'scripts' to provide a framework for pack design, we first need to understand the key elements that combine to influence brand choice (Fig. 7.1).

The optimum balance between the 'rational' and the 'emotional' elements shown in Fig. 7.1 will differ according to the relevant 'script' (Fig. 7.2). In addition to the overall balance between rational and emotional

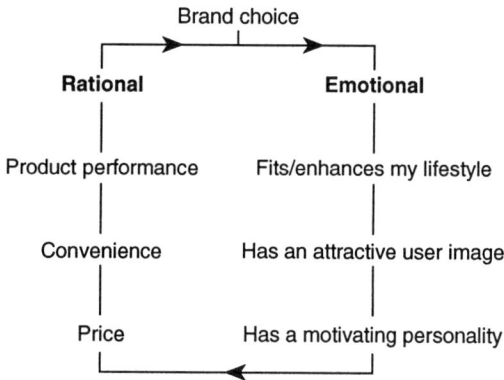

Figure 7.1

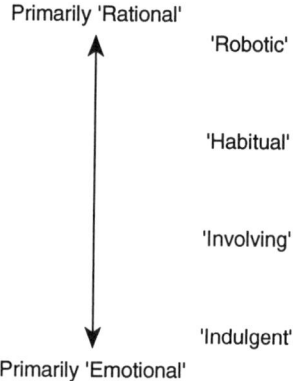

Figure 7.2

Table 7.1 How packaging takes account of communication requirements

Communication requirements	Language components
Recognition	'Anchor Points' define the brand, and are particularly important when evolving existing brands and introducing line extensions.
Understand	'Market Codes' communicate product type. New products often need to use the market's existing visual language in order to generate understanding and fit into a 'script'.
Want/Tempt/Desire	'Key Triggers' convey relevant emotional brand benefits. In some markets, the key triggers can be promotion driven.

Table 7.2 How language components work

Language component	Characteristics
Anchor Points	Very fast communication, which needs to work at 5–6 m distance in a large supermarket.
Market Codes	Quick communication – closer but still needs to communicate at a distance of 2–3 m.
Key Triggers	Close communication allowing time to absorb/enjoy, a maximum of 1 m.
Information	Studied detail – communicates at reading distance.

attributes, it is essential that the pack takes account of specific communication requirements by exploiting the pack's 'language components' (Table 7.1).

Each 'language component' needs to work at a different distance in order to match the speed of recognition of the different levels of communication (Table 7.2). Like all good theories, this one conforms to common sense principles, and can be tested through observation.

7.2.2 Who makes the purchase?

It is important to distinguish between different types of purchase, and whether they are usually undertaken by the child on its own or amongst others. A child on its own can be observed behaving very differently from a child in a group.

Watch and observe children buying for themselves in toy shops, and you will immediately observe the visual 'triggers' in action.

Pocket money purchases come top of the child's desirable purchases – and fall into the 'indulgent' category in our classification of fast-moving consumer goods. Toys are, of course, the most popular items, and Saturday the most popular shopping day. Taking time to consider the purchase is an essential part of the enjoyment and is where the packaging plays such a vital role.

Children will browse in toy shops looking to be motivated, like adults browse in book shops. The pack provides a number of key 'triggers' including information about the product, ideas of how to use it, with play suggestions or settings, like 'Cindy', 'Lego' or 'Playmobil'.

Contrast this with packaging for toys which are essentially bought by other people for children. 'Tomy' toys are designed to achieve shelf impact and fit the category language that defines the age group. This is unnecessary when children are purchasing for themselves.

If children are not in a position to design their own packaging what would they like to see others do to appeal to them? We commissioned some fascinating research aimed at trying to answer this question.

7.3   Designing packaging to appeal to 6–9-year-olds

The methodology used was focus groups, targeting 6–9-year-olds, the age when children start to enjoy self-purchase independence and, with it, develop strong views about branding. Boys and girls were researched separately in a study to identify this group's attitudes to packaging design, with a particular focus on confectionery, savoury snacks, biscuits, soft drinks, yoghurts and cereals.

The following represent the topics which emerged from the children's observations as key factors to take into account when designing packaging to appeal to children.

- The design elements that tended to attract their attention first.
- Their favourite colours or colour combinations and the least attractive ones. Differences emerged between boys and girls.
- Their attitudes to flavour colour codes, typography, illustration (realistic and abstract) and product information.
- The importance of product names and their awareness of manufacturers' names or brand identities.
- The kind of promotions they respond to.
- The types of characters which appeal to them and why. Manufacturers' own characters created for a brand compared with licensed characters. Short-term fashionable characters, versus 'Disney', 'Tom and Jerry' and other long established characters.
- Characters, in general, versus appetizing messages with photography or illustration of real ingredients/products.
- Characters, versus a more adult imagery – aspirational brands like 'Coca Cola'.
- The extent to which the products' flavours, their ingredients or their characteristics are important motivators.

Whilst it may be misleading to generalize from one specific piece of research, some interesting general observations emerged.

7.3.1 Colours

Their favourite colours were (a) purple, red and yellow and (b) blue and green. Orange and pink produced polarized reactions. In general, the children tended to dislike light, dark, 'dirty' and sophisticated colours. Large colour blocks were perceived as boring. Foil, although identified as 'bad for the environment', helped to achieve impact and appeal.

The children responded to colour movements and bold colour contrasts. Generally speaking, colour contrasts allowed brands to use more 'adult' background colours. Colour contrasts also enabled brands to use colour codes outside their sector's visual language, for example 'Munch Bunch'.

Flavour colour coding was core to their understanding of the product proposition. These must be very clear, as they strongly influence their choice and increased 'pester power'. Colours which cause confusion over flavour should be avoided, for instance, those which appear on 'Fiendish Faces' and 'Petit Filous'.

7.3.2 Multi-packs

Multi-packs were perceived as environmentally unsound. They could also be very confusing, preventing children from understanding the product flavour mixes. They preferred to see the products inside, for instance, the 'Fruit-tella' transparent multi-packs.

7.3.3 Typography

Bubble writing was their favourite typography, for example, Rowntree's 'Fruit Juice' and 'Happy Faces'. Refined and straight typographies, such as 'Ribena' are perceived as cold and unfriendly.

7.3.4 Illustration

The more innovative logotypes, using different colours, for instance, 'Dinosaurs' or product illustration, like 'Honey Nut Loops', scored on impact and appeal.

Realistic illustrations did not add much appeal to the products; they could be perceived as adult and, therefore, boring. 'Petit Filous' was regarded in this way.

The children unanimously liked the illustrations on Rowntree's 'Fruit Juice Drink', because of their colours, their fun and playful looks and their prominence all over the pack.

Illustration was preferred to photography. To be credible, illustrations should adopt the same visual language as the rest of the pack design. A good example is the 'Wildlife' pack from Yoplait.

Appetite appeal did not feature as an important issue, except in the case of chocolate illustrations, for instance 'Happy Faces' and 'Petit Filous'.

7.3.5   Use of characters

The appealing characters were those intimately linked with the products. The most successful ones were either connected with the product's flavours – 'Munch Bunch' – or its physical attributes – 'Happy Faces'. Equally, the characters having a direct interaction with the product – McVitie's 'Tom and Jerry', or the characters at the core of the brand proposition –'Monster Munch' – performed very well.

Packaging and advertising can work very well in synergy. Any character used in advertising should also be credible on packs. This was a problem which 'Robinsons' encountered until design changes were made. Indeed, packaging bridges the gap between the inaccessible world of dreams and the realm of reality.

The older children were cynical about licensed characters. Licensed characters, if used as part of the brand proposition, should be clearly differentiated from promotional offers; for instance, there were problems with the identity of Cross & Blackwell's 'Fun Pasta'. Relevance, credibility and longevity were the main pitfalls pointed out by the children themselves.

The popular appeal of characters clearly drops off rapidly with the older age group in our range of 6–9-year-olds. One wonders long term what the effect on the category will be of HP's 'Postman Pat', 'Sonic the Hedgehog' and 'Mr Blobby', confronting Heinz's 'Thomas the Tank Engine', 'Noddy' and 'Super Mario'!

7.3.6   Conclusion

In general, the packs which were most favoured by the children in the focus groups, combined the following characteristics:

- impact (brightness, speed and movement);
- a friendly and inviting personality and the possibility of a long-term relationship with the product;
- a lot of detail to examine and dream about;
- a relevant proposition that is not 'babyish' or patronizing;
- honesty (a clear product proposition, flavour coding, genuine and meaningful characters);
- focus on taste delivery (fruitiness, chocolate appeal);
- design and advertising working in synergy.

7.4 Creating packaging to last

One fundamental question which must be answered is if children change rapidly, as they do, is it necessary to keep changing the packaging? How much must packaging be a dedicated follower of children's fashion? Which are the all-time favourites, and how are they created?

Character merchandizing is, by definition, short term, and its function is entirely clear and, in fact, transparent as far as children are concerned. However, old favourites have particular appeal to the parent and can make an unexpected come-back, for instance, 'Noddy'.

Characters which are created for the product and form an integral part of the proposition have great potential strength, and if this forms part of the advertising and promotional message, the message is reinforced – as the well-exploited 'Milky Bar' kid will testify.

The other way characters can be effective with children, is to borrow a brand from one category and stretch it into another, such as Bassett's 'Jelly Babies' appearing in the chilled cabinet. But enough has been written about brand stretching in other places without it being necessary to cover the same ground here.

7.5 Older children

However, appealing to the older age group of children, even from eight to nine upwards, takes a very different approach. Here, brands such as 'NRG' and 'Capri Sun' and 'Frijj' score highly. The target group is aspirational, and looking to brands to provide a more sophisticated proposition.

Children from a younger and younger age are looking increasingly for more adult values, total branded propositions which convey a message about the purchaser/consumer and the relevance of the product to meet their needs.

Acknowledgment

The data for this chapter was originated by P.I. Design International.

8 How much does food and drink advertising influence children's diet?

T.P. BARWISE

8.1 Introduction

In this chapter, I explore the influence of advertising on children's diet. This is of more than academic interest. As in the USA 20 years ago, there is now in Europe a lobby arguing that children's food and drink advertising has a strong and unhealthy influence on what young people eat and should be banned or severely curtailed. This lobby represents a real threat to manufacturers' freedom to develop and market their products.

In the following sections, I first review the anti-advertising case, then summarize some general arguments on behalf of the advertisers, before discussing in more detail the various influences on children's diet and, specifically, the role of advertising. The chapter concludes with a brief discussion.

8.2 The anti-advertising case

The case against advertising children's food and drink rests on three main premises:

1. **Children's food and drink advertising is overwhelmingly for processed products, mostly from a limited number of 'unhealthy' categories; there is little advertising for 'unprocessed' foods, such as fresh fruit:** 'Ninety percent of child oriented advertising is via television . . . Food adverts on children's television are dominated by pre-sweetened cereals, confectionery, fast foods, snacks and soft drinks. Advertising of such a narrow range of products cannot be said to encourage healthy choices or promote a varied and healthy diet' (Dibb, 1993, pp. 1 and 14–17).
2. **Advertising has a marked effect on what children eat:** 'Advertising not only directly influences children's food preferences and choices but also does so indirectly through its influence on parents and peers . . . Children are more responsive to and influenced by advertising than adults . . . Their "pester power" influences . . . the family shopping . . . Young children may lack the skills to assess let alone understand advertising's purpose . . . Advertising, if so designed, could help children choose a healthier diet.' (Dibb, 1993, pp. 1, 13 and 18–21).

3. **The food children eat affects their health both as children and as adults:** '[Bad] diets . . . are not only a significant factor in the development of childhood health problems . . . but also set the pattern for the development of [serious diseases] . . . in later life.' (Dibb, 1993, pp. 1 and 5–7).

The first of these three premises is not in dispute. Television is usually the most cost-effective medium for attracting children's attention, partly because the cost per viewer is typically lower than for adult audiences, because of children's much lower spending power. The food and drink categories advertised to children are those they like and buy or help to choose (e.g. children rarely buy breakfast cereals but strongly influence brand choice). Advertising is for products which can be branded and differentiated: 'unprocessed' products like fresh fruit are largely unbranded commodities, not only because of the difficulty of differentiating them from competitors' products but also because of the problem of ensuring consistent quality and quantity of supply.

The second premise is the subject of this chapter. Briefly, I argue that anti-advertising lobbyists have a vastly exaggerated view of the influence of brand advertising on children's consumption of the relevant product categories.

The third premise involves a cluster of issues linking diet, nutrition, lifestyle and health. Diet can clearly be a factor in the development of childhood health problems. Similarly, childhood diet can have longer-term health effects both directly (via long-term effects on physiology) and indirectly (via long-term habits and preferences). How significant these effects are, in practice, is an empirical question, e.g. the extent to which childhood health problems in Britain are diet-related.

Before addressing the second premise – the main focus of this chapter – I briefly review the advertisers' broader case, including their disagreement with the anti-advertisers' implicit assumption that the food products advertised to children are, in themselves, unhealthy.

8.3 The advertisers' case

Leaving aside the question of how food and drink advertising influences children's diet, four arguments can be used for allowing manufacturers to advertise to children – subject, of course, to the normal consumer protection regulations for products and advertisements.

1. *Freedom of speech.* The first argument is a question of principle – free speech. In the words of the political philosopher John Gray, 'One of the most characteristic features of our age is that, as artistic and cultural freedoms have expanded, freedom of commercial expression has

tended to shrink. At the same time that playwrights, novelists, and film-makers reject any restraint on their freedom to challenge established mores and beliefs, advertisers find themselves subject to ever more restrictive regulations regarding which products they may promote, and the conditions of their promotion' (1991, p. 9).

Advertisers agree with the need for consumer protection, i.e. that advertisements should be legal, decent, honest and truthful (while wryly noting that these restrictions do not apply to political advertising). They do not seek the freedom to make claims like 'With its unique power to stimulate cellular activity, [Bovril] strengthens the body against germs. It has been shown to possess body-building capacity equal to between ten and 20 times its own weight' (quoted by Girling, 1995, p. 26). Subject to these necessary controls, however, they seek the freedom to promote any legal product.

Although freedom of speech may sound like an abstract, philosophical principle, its embodiment in the First Amendment to the US Constitution gives it great influence there. It was a major reason for Congress's rejection of the US Federal Trade Commission's proposed restrictions on television advertising to children in the late 1970s (Kunkel and Roberts, 1991, pp. 65–6).

2. *Paying for children's programmes*. A second argument largely ignored in Europe is that about three-quarters of the money spent on advertising to children is on airtime costs (as opposed to agency fees and commercial production costs). Airtime costs go to broadcasters and pay for children's programming. The importance of this argument is that it shows that advertising restrictions are not, as they might appear, costless.

In a system where programme content is largely deregulated (as in the USA, Italy, etc and now the UK), if broadcasters receive no revenue for showing children's programmes they will stop doing so. This tradeoff was explicitly recognized in the Federal Communication Commission's (FCC) 1974 limits of 9.5 minutes of advertising per hour of children's programmes at weekends and 12 minutes per hour on weekdays (Kunkel and Roberts, 1991, p. 61). The FCC accepted the broadcasters' argument that they needed this advertising to fund the programmes. The lower time limit at weekends reflected the higher audience ratings and therefore revenues for advertising slots on Saturday mornings.

Even in a more regulated system where broadcasters can be required to continue showing children's programmes with no direct revenue to pay for them, the effect is to divert funds from other (i.e. adult) programming. The advertising restrictions still have to be paid for by viewers, although not only by child viewers.

3. *Advertised products are not 'unhealthy'*. Thirdly, food and drink

manufacturers strongly disagree with the presumption that their products are in themselves 'unhealthy'. The emphasis in modern nutrition is on healthy diets not on individual foods in isolation. Of course it would be unhealthy for a child to subsist on nothing but, say, potato crisps. But crisps are a normal part of a healthy, balanced diet for young people. In the words of Dr Tom Sanders, Reader in Nutrition at King's College, London, 'Crips account for less than 2 percent of a child's energy requirement but supply between a quarter and a third of the Vitamin E intake and a third of Vitamin C. There is twice as much Vitamin C in a packet of crisps as there is in an average apple' (quoted by Girling, 1995, p. 29).

Sanders also believes that the National Food Alliance (NFA) is . . . 'wrong about sugar causing heart disease and obesity. Recent studies actually show that people who eat more sugar are thinner. And the possibility is that people who cut out sugar eat more fat. If kids are getting fatter, the reason is lack of exercise' (quoted by Girling, 1995). Other nutritionists also emphasize lack of exercise: 'It is generally assumed that ready access to highly palatable food induces excess consumption and that obesity is caused by simple gluttony. However, average recorded energy intake in Britain has declined substantially as obesity rates have escalated. The implication is that levels of physical activity, and hence energy needs, have declined even faster', according to Andrew Prentice and Susan Jebb (1995) of the Medical Research Council's Dunn Research Centre, Cambridge. Similarly, Fox (1993) concluded that children and teenagers may be less fit than before because of reduced levels of exercise, such as school sport.

Sanders believes that the lobbyists' constant negativity – their insistence that certain foods are 'bad' – has induced unnecessary and harmful anxiety among young people: 'The thing most likely to kill an adolescent girl is *not* eating. And we are seeing it in boys now. They have these obsessions that they can't eat any fat or any sugar' (Girling, 1995).

4. *Long-term diet and health trends.* Finally there is evidence of long-term improvement in children's diet and health, despite several lifestyle trends which go the other way, such as fewer school meals and formal meals at home, less exercise, and more opportunity for children to decide what they eat (with nutrition rarely an important factor in their choice). Moreover, insofar as advertising has had a role in launching new, 'healthier' products, it has contributed to this improvement.

The anti-advertising lobby regards most advertised products as 'unhealthy' and treats their consumption as virtual proof of poor or even declining diet and health. In particular, the NFA report (Dibb, 1993) included a number of misleading statements about children's health, diet and nutrient deficiencies. I here repeat the relevant points from my commentary (Barwise, 1994).

8.3.1   Children's health

The report noted (p. 5), that 'Growth and stature are common measures of overall nutritional status'. It omitted to say that by this criterion UK children have never been healthier. Since World War Two, they have been growing steadily taller and are now taller than ever before. One of the NFA report's own references – ref. 5 on the same page – found that 'children of both sexes aged 5 to 11 were taller in 1981 than in 1972' (Rona and Chinn, 1984).

Dental health has also enormously improved. The dmf index (decayed, missing, filled teeth) for British 5-year-olds declined from 5.7 in 1958 (HMSO, 1960) to 1.86 in 1988 (Dowell and Evans, 1989).

8.3.2   Children's diet

It will hardly surprise readers that 'children's consumption of foods such as confectionery, biscuits, soft drinks, crisps, chips and burgers is far higher than the average for the whole population' (Dibb, 1993, p. 11). Anyone with children will know that these are the kinds of foods they tend to choose if given a choice. (The various influences on such choice are discussed later in this chapter). As the NFA report's Table 3 (p. 11) also shows, other foods eaten disproportionately by children include breakfast cereals, pizzas, fish fingers and pasta, partly reflecting convenience of preparation.

The question, however, is not how children's diet compares with adults', but whether (and if so, how) it differs from a balanced diet which research suggests is healthy for children.

Most nutritionists would agree that most children's diet would be improved by including more vegetables and fresh fruit. The NFA report notes that the Consumers' Association (1992) found that more than 25% of schoolchildren had no fruit (or even fruit juice) or vegetables over a one-day period. However, saying that children should eat more fruit and vegetables is not at all the same as saying that they should eat fewer advertised foods. The NFA report provides no evidence whatsoever that the foods and drinks advertised on children's television should not form part of a healthy balanced diet for children.

8.3.3   Slimming and micronutrient deficiencies

The report (p. 6) noted that 'slimming is an important health problem in adolescence'; that over half of all British 16–19-year-old girls have been found to have been on a diet at some stage to lose weight (Health Education Authority, 1990b); and that teenagers on slimming diets are more likely to be short of essential vitamins and minerals, especially iron. Referring to a study by Nelson, White and Rhodes (1992), it said:

'Anaemia was three times more common in girls than in boys and was particularly common among girls who reported that they had tried to lose weight over the past year (23%) and among those [on] vegetarian diets (25%)' (Dibb, 1993, p. 6).

It also said that there has been 'much concern about whether diet can affect intellectual performance' at least for 'a small minority of children in Britain who have sub-clinical vitamin and mineral deficiencies' (p. 6), referring to a short review by Addy (1986). It did not say that Addy's review was specifically about iron deficiency and concluded that the evidence to date would not justify a national campaign of extra iron supplementation. Nor does it mention that the one case of wider deficienies referred to by Addy (vitamin D, energy and iron) was for children with Bangladeshi parents, whose diet presumably reflected their families' culture. Addy suggested that 'The size of the problem should be assessed by community based surveys of children from various racial, cultural, and social backgrounds' (Addy, 1986; Sharp, 1994).

The NFA report also referred to evidence of some calcium and vitamin A deficiency among teenage girls and, to a lesser extent, boys (p. 8) and that 9% of children have nothing to eat before they leave for school each morning (p. 10).

It apparently saw no contradiction between these findings and concerns and its main thesis that (a) British children eat too much of the food advertised on television and (b) a major reason is that these foods are advertised. In particular, it repeatedly criticized the heavy advertising of pre-sweetened cereals on children's television. But, if advertising were as powerful as the NFA claims, children and teenagers would be 'forced' to start the day with cereals (fortified with vitamins and iron) and (calcium-rich) milk instead of going to school with no breakfast. This would eliminate, or significantly reduce, these micronutrient deficiencies.

Finally, the report's claim that figures from an earlier study by Nelson (1991) suggest that 'micronutrient intakes may also be falling below adequate levels' (p. 8) significantly misrepresents what Nelson actually wrote. The original paper gave this statement in its introduction as a 'notion' to be tested empirically. In the event, the findings were that:

- 'intakes were substantially above the RDA for all nutrients except energy, Fe, and vitamin D',
- 'in spite of an apparent imbalance in the food profile of these children's diets, one cannot say that nutrient adequacy (apart from Fe) or body-weight was being adversely affected'.

● 'we must therefore conclude that the diets of these children, although imbalanced in terms of current guidelines, and possibly in terms of long term health, were of sufficient quality not to impair intellectual performance' (Nelson, 1991).

8.3.4 Overall food consumption trends

Most evidence relating to the dietary habits of children is derived from small-scale surveys. Far more evidence of a much higher quality is available relating to the food consumption of the British population as a whole, although even this is largely limited to expenditure on foods purchased for consumption in the home. It is clear from government statistics that the in-home eating habits of the population at large have changed very significantly over the last 25 years in the general direction favoured by government dietary recommendations. For example, consumer expenditure on red meat, many dairy products and white bread has fallen sharply, in real terms. In contrast, consumers have spent more on products recommended by government agencies as 'healthier', such as fresh fruit, vegetables, fish and poultry (Advertising Association, 1993).

In addition to the basic consumption statistics, large-scale commercial surveys also show that consumers' attitudes to a whole range of foods have changed significantly in recent years, as would be expected from the change in consumption patterns.

Finally, advertising expenditure patterns have also changed markedly as a result of changing consumer attitudes and preferences. Advertising for most foods the NFA regards as 'healthy' is sharply up. Advertising for most foods it regards as 'unhealthy' has fallen. Although the definition of what is 'healthy' and what is 'unhealthy' can vary widely (through time as well as between sources), UK advertising for 'healthy' products increased in real terms by over 126% between 1975 and 1992, the advertising of 'neutral' products by over 67%, and the advertising of 'unhealthy' products by 20% (Advertising Association, 1993). Making allowance for the increased cost of commercial TV airtime, the advertising of 'unhealthy' products has fallen significantly in impact terms over the last 20 years.

Unfortunately, the government statistics do not cover food consumption specifically by children, nor consumption outside the home. The indirect evidence (e.g. from data on height and weight) is that children's nutrition has improved significantly over the 35 years since the advent of commercial television. To my knowledge, the only direct evidence from repeated cross-sectional studies in the UK found little change in the diet of 14-year-olds in Glasgow over the period 1964–71 (Durnin et al., 1974) and in that of 12-year-olds in Northumberland over the period 1980–90 (Adamson et al., 1992). The former study did find an increase in body fatness among boys, associated with a marked reduction in physical activity.

8.4   Influences on children's food choices

Like other anti-advertising lobbyists, the NFA report assumes that, relative to other factors, advertising is a major influence on children's food choices. It devotes little space to examining any evidence or arguments for or against this assumption, perhaps regarding it as self-evident.

However, in reality there is no evidence that advertising is a major influence on children's food choices; at the same time, there is substantial evidence that it is not a major influence, and that other factors – notably parents – are a much stronger influence.

The report says that 'it is not within the scope of this report to examine this area [i.e. the influences on children's food choices] in depth' (p. 12). It, therefore, devotes just a page and a half to a brief and largely uncritical review of six influences:

● taste preferences;
● exposure to and availability of food;
● parental influence;
● peer group influence;
● knowledge, attitudes and beliefs; and
● advertising.

The tone of this discussion is that advertising is 'an almost universal influence' (p. 12). This is reflected in the report's claim that advertising is not only a direct influence but also influences children's food choices indirectly under all of the first 5 influences listed above. According to the report:

● taste preferences can be changed by videos showing role models eating unpopular food such as broccoli, spinach and fruit (Nuttall, 1992). This latter reference is a newspaper report of some recent work at Bangor University on short-term responses under experimental conditions. However, longer-term real-world evidence is that 'years of watching and enjoying "Popeye" cartoons, in which this much admired central character became strong, heroic and successful whenever he ate spinach, failed to persuade the children of America that they liked and wanted this vegetable' (Esserman, 1981);
● exposure to television advertising is not only 'almost universal' but also presents the advertised foods as 'desirable and attractive' (p. 13), which also shapes food choices (Rousseau, 1984);
● parents and peers are themselves exposed to and influenced by advertising;
● children are more receptive to messages conveyed by advertising than by classroom lessons or textbooks (Iceland, 1990); in any case, there is little correlation between children's (often well-developed) knowledge about

what is healthy and their actual food choices (Gardner Merchant, 1990; Leatherhead Food Research Association, 1991).

In contrast, the direct influence of parental control is covered in the eleven words 'Parents control <u>to some degree</u> the food eaten in the home' (underline added).

8.4.1   Product categories versus brands

The NFA report considers that advertising has a big direct influence on food choice:

'Studies demonstrate that advertising increases the desire for and purchase of the advertised product [*no references*]. This is hardly surprising as it is, after all, the purpose of advertising. Companies would not spend large sums of money on advertising if it did not have a positive effect on sales' (p. 13).

However, this confuses **product categories**, like soft drinks, with **brands**, like Sprite. This elementary distinction is crucial in this context (Gardner and Levy, 1955; King, 1973).

Nutritionists' concern is with the way children's diet is allocated across broad categories like confectionery, crisps, soft drinks, fruit, vegetables and cereal. It is not about whether the confectionery is Snickers or Aero, or the oranges are Outspan, Jaffa or Spania. But this is exactly what concerns advertisers and is in most cases by far the main reason they advertise, especially in mature markets like most food categories, the only other important reason being to support a price premium against discount and private label brands (Yasin, 1995). As a general statement, advertising has little or no effect on **total category sales** (Simon, 1970; Ehrenberg, 1974; Lambin, 1976; Esserman, 1981; Henry, 1984; the recent reviews by McDonald, 1992 and Walters and Ambler, 1993).

This distinction between product categories and brands is pretty clear cut, but needs some qualification. The most important relates to category definition and product substitution.

1. Categories can be defined broadly (e.g. soft drinks) or narrowly (e.g. diet lemon-lime carbonated soft drinks). Heavy advertising of a narrowly-defined 'category' often leads to some 'category' growth. Feldwick (1995, p. 65) gives the following figures:

   • 'One packaged goods category, we estimated, declined by 13% over five years as a result of the market leader cutting back on advertising.
   • In another, advertising adds an estimated 3% a year to the category size.
   • In a household product category, £1m of advertising increases category sales by £4m.

These are examples of pure "brand" advertising. In so called "generic" campaigns results can be more dramatic:

- Sales of one basic fresh food category up 12% in one year when advertising started.
- Another, major staple item where advertising added 6% to volumes over a decade.'

'. . . On the positive effects of advertising on category size we might cite the example of the stock cube market in 1978 when the Bovril cube was launched as a rival to Oxo. Both brands spent huge sums in a battle for position – as a result, this extremely mature market increased 20% in size year on year'.

Similarly, Goldberg (1990) found – admittedly, only on a cross-sectional basis – a positive relationship between the viewing of American children's TV by both English- and French-speaking children in Quebec and the number of different children's cereals in the home, after a Quebec law banning advertising on the local (French-language) stations.

In general, we would expect such 'category' growth to be largely at the expense of near-substitutes in the wider category, although this has not to my knowledge been the subject of any systematic research as it is of little interest to marketers. Generally, however, the whole basis of competitive positioning in marketing (Ries and Trout, 1985) – the rationale behind product strategy and advertising, and to some extent pricing, promotion and distribution – is that competition is most intense between products or services perceived by the consumer as most similar and substitutable. What constitutes a near-substitute is therefore determined by consumer perceptions; measuring these is an important part of market research (e.g. Lunn, 1986; Mercer, 1992, ch. 6; Kotler, 1994, chs 9, 11 and 12). Economists define the degree of substitutability as the 'cross-price elasticity', i.e. the extent to which the demand for X is influenced by a 1% change in the price of Y (Chamberlin, 1965; Porter, 1976; Lipsey, 1983). Because causal cross-price elasticities are hard to estimate, recent demand theory incorporates consumer perceptions which are more measurable (Lancaster, 1979; Dickson and Ginter, 1987).

The key point in the current context is that any growth in a narrow 'category' (diet lemon-lime carbonated soft drinks) caused by heavy advertising is likely to have little, if any, effect on nutrition, except insofar as some brands and subcategories have, say, more sugar than others, e.g. pre-sweetened versus unsweetened cereals (although most people add their own sugar to the latter). However, in this case flavour and texture differences are likely to be a much stronger influence on choice between subcategories than advertising. Thus, advertising a chocolate biscuit brand may attract some sales from other rather similar sweet snacks as well as from directly competing brands. The perceptual data suggest that it would be unlikely to attract sales from oranges or brown rice, or even from muesli, yogurt or bananas. Even here, however, there can be exceptions:

> 'Another anecdote . . . concerns a large food category which has been in decline for several years. A very thorough analysis showed that the decline began about two years after the market leader stopped advertising; that switching was not to other, substitutable products, but to eating nothing at all on those occasions when the product was used; and that none of the other hypotheses for the market decline stood up upon examination.' (Feldwick, 1995, p. 65).

Other qualifications to the general statement that advertising has little or no effect on total category sales are that:

2. Advertising can accelerate total category sales during the 'early growth' phase. The category may be fairly new (e.g. camcorders) or with recent product improvements (e.g. instant tea, rechargeable batteries) (Robertson, 1971; Rogers, 1983; Golder and Tellis, 1993).

3. Anecdotal evidence suggests that, for products that consumers tend (or prefer) to 'forget', such as DIY products, advertising can increase category sales by acting as a reminder of the product and its use. A recent UK case of successful 'reminder' advertising for a service consumers take for granted was the so-called 'dancing bottles' campaign for doorstep delivery of milk (Baker, 1993).

4. Advertising is important for the launch of new brands, brand extensions, and line extensions in an existing category (e.g. a new flavour of soup, a new model of car, see King, 1973; Aaker, 1991). If the new product is strongly differentiated, its launch can lead to category growth (e.g. the entry of Mars and Häagen Dazs into premium ice cream). But this will not be sustained unless the products have real differential value to consumers (Davidson, 1972; King, 1973).

5. If one supplier has a virtual monopoly, its advertising will, if successful, increase total category sales (e.g. BT call stimulation). This applies to none of the categories discussed in the NFA report, although it does apply to some subcategories as in (1) above.

6. Occasionally, trade associations or the like pay for 'generic' advertising for a whole category, like oranges or meat. The effects on category sales are generally small, although recent research suggests that producers may be able to obtain higher prices as a result of the advertising (Green, Carman and McManus, 1991; Jensen and Schroeter, 1992; Zidack, Kinnucan and Hatch, 1992). However, one recent study of the French

market for salmon (Bjorndal, Salvares and Andreasson, 1992) and one of milk in Canada (Venkateswaran and Kinnucan, 1990) provide tentative evidence of generic advertising leading to increases in total consumption – subject again to the question of product substitution discussed under (1) above.

These six provisos do not materially affect the main point that **nutrition** is almost entirely about **product categories** while **advertising** is almost entirely about **brands**.

The NFA report does mention this argument briefly later (p. 20) but then dismisses it in one paragraph, citing a single study (about tobacco). It also argues that the fact that 'the market for child-orientated products is said to be buoyant' proves that brand loyalty affects total consumption. This appears to be a *non sequitur*, again reflecting the failure to distinguish brands from product categories. There are many ways of defining and mesuring brand loyalty (Jacoby and Chestnut, 1978; Engel, Blackwell and Miniard, 1990). Within each category, all of these measures tend to vary predictably with market share (Ehrenberg, Goodhardt and Barwise, 1990). However, there is no necessary or even probable relationship between any of these measures of loyalty for individual brands and the size or growth of the total category. Nor, to my knowledge, has there been any empirical research or even theoretical conjecture about any such relationship.

8.4.2   Influences on children's choice of food categories

I am not aware of any major systematic research on the influences on children's choice of broad food categories. It would be an extremely difficult topic to research comprehensively, requiring, in my view, a combination of qualitative and quantitative methods and some longitudinal studies. The current MAFF study should throw light on the problem, but it remains complex.

Nevertheless, I believe we have enough evidence to assess the main factors.

1. Taste preferences: children have innate taste preferences which vary across individuals but also systematically differ from those of adults. For instance, the NFA report (p. 12) notes that

   'New-born babies have been shown to prefer sugar solution to water . . . and older children have been shown to have a continuing preference for sweet tastes.' (Thomas, 1991).

   Children and teenagers also like many savoury foods such as pizza (Guber and Berry, 1993).

2. Parents: children's familiarity with and preference for different types of food (mainly categories, but also brands) is strongly influenced by

parents. This influence includes both direct control of what is consumed in the home (especially for younger children) and, as the NFA report notes, the strong influence of parents' own attitudes and behaviour towards different foods.

3. Food provided outside the home: even outside the home, much of what teenagers and, especially, children eat is provided for them. This especially holds for school meals and meals at friends' homes, both of which are largely chosen by adults. It also includes 'eating-out', where children have more influence on choice (McNeal, 1992; Guber and Berry, 1993), with taste the dominant criterion.

4. Other influences on children's attitudes and behaviour: from an early age, some of what children eat is bought by themselves out of pocket money and, increasingly as they get older, dinner money (McNeal, 1992; Guber and Berry, 1993). Their main criterion is again probably taste, but convenience, availability, and value for money are also important. Children are also influenced by attractive display and packaging. (My personal experience suggests that they are unwilling to pay a significant price premium for display and packaging if it has to come out of their own pocket, but I have seen no published research on this.) Their attitudes and beliefs are also influenced by peers, teachers and the media (not just television, not just advertising, and certainly not just advertising on children's television). The influence of peers is hard to separate empirically from that of parents, siblings, and peers' parents, all of whom tend to come from the same social milieu. The influence of teachers may, as the NFA report suggests, be fairly limited, at least in the short term: knowledge of nutrition has little effect on food choice (Gardner Merchant, 1990; Iceland, 1990; Leatherhead Food Research Association, 1991; Guber and Berry, 1993). The influence of advertising on **category** (as opposed to **brand**) choice appears to be very limited. However, media portrayals (including in advertising) may reinforce the desire to be slim, especially among adolescent girls.

8.4.3 The relative importance of these influences

The influence of individuals' innate **taste preferences** is not in dispute. As with most 'nature–nurture' issues, it is virtually impossible to isolate the influence of this factor. It should be stressed, however, that commercial success in packaged goods is determined by repeat-buying which, in the long run, depends largely on consumer taste preferences.

The evidence is that parents are overwhelmingly the strongest environmental influence on children's developing food habits and preferences. The ECDU (1987) study's findings indicate that by the time Welsh children were eighteen months old they were eating very much like their parents.

The NFA report also (p. 12) refers to Ritchey and Olsen's (1983) finding that children's frequency of eating sweets was greatly influenced by parents' own frequency of sweet-eating. Ritchey and Olsen also found that parents' 'positive attitudes toward giving their children sweets in positive contexts' were another strong influence. The report also notes that Escalona's (1945) classic study of very young children under experimental conditions found preferences for orange or tomato juice directly influenced by the parent's choice.

More generally, the enormous differences in food habits between countries and cultures, and (within countries) between different regions, social classes and ethnic and religious groups, all reflect the cultural significance of food. Food preparation and consumption play a major role in the everyday life of all societies (Douglas, 1984; Mars and Mars, 1993).

In modern (especially non-Mediterranean) Western societies, these everyday patterns are mainly transmitted through the nuclear – or post-nuclear – family. For various reasons (more working mothers, freezers, microwaves, VCRs, cars, disposable income growth, etc.) the steady trend has been towards convenience, informality, and 'grazing'. There are also marked differences between families in both eating habits and styles of parenting and socialization, related to variations in the ordering of household relationships generally (Mars and Mars, 1993).

In the last 10–20 years there has also been a change in the food provided outside the home, namely the sharp reduction in the provision of school meals from 4.9 million UK children in 1979 to 2.8 million in 1988, the last time comparable data were collected (Cole-Hamilton, Dibb and O'Rourke, 1991). Unlike advertising, this factor (a) directly affects what children eat and (b) has changed significantly in recent years. Given Dibb's involvement in the School Meals Campaign, it is remarkable that she does not mention this important factor as an influence on children's eating patterns, especially for poorer children.

Among other influences, we have already noted that children's knowledge of nutrition has little effect on food choice. The NFA report (p. 18) cites Atkin's (1975) finding – which will hardly come as a surprise to most parents – that 'in more than 99% of cases children did not make any explicit mention of either vitamins, minerals or the general health value of the product in choosing a cereal [in the supermarket]'.

This does not mean that knowledge has no influence at all. As already noted, over the long term, there have been significant shifts in consumption patterns, largely driven by consumer knowledge and the increasing availability of 'healthy' foods. On a smaller scale, All-Bran's recent advertising has increased sales by communicating how much fibre it contains (Baker, 1993). Similarly, the hostility of today's young adults towards drink-driving reflects their knowledge as well as their values (Portman Group, 1993). But such knowledge must be communicated thoughtfully and skilfully, and not in the expectation of producing dramatic short-term change.

8.4.4   Advertising

The NFA report stresses the universality of television advertising without apparently realizing that this very universality would be inconsistent with the enormous observed differences in food habits between cultures, subcultures, families and individuals if advertising were as strong an influence as the report assumes. We have already noted that pre-sweetened cereal advertising (the volume of which is criticized by the NFA report) does not deter children from missing breakfast or girls from going on a slimming diet. Instead, these advertisements aim to influence brand choice among those who do eat cereal for breakfast. The Quebec study by Goldberg (1990) already mentioned suggested that heavy viewing of American children's TV may increase the number of children's cereal brands in the home. It did not – unfortunately – go further to suggest that such viewing actually stops adolescents from skipping breakfast.

Other strong evidence that television advertising has only a limited influence on children's preferences for food categories is that when they are free to buy what they want, children's favourite choices include some virtually unadvertised categories like take-away chips and (for smaller children) unwrapped sugar confectionery. Conversely, categories they do not choose include some like coffee and beer which are heavily advertised and include brands whose advertising they like. However, liking the brand's advertising has little to do with consuming the product category. A 1990 survey found that, among UK children aged 7 to 14, their top three favourite commercials were for Oxo stock cubes, Andrex toilet tissue and Weetabix, respectively (BTHA, 1990). Other favourites included BT, British Gas, PG Tips tea and Persil detergent.

These are broad arguments. We now consider the evidence on advertising in more detail.

8.5   Children and food advertising

8.5.1   The effects of advertising on children's food preferences

There is a massive literature on the effects of advertising in general. This is the main topic covered in refereed journals such as the *Journal of Advertising Research* and the *International Journal of Advertising*. It is a major topic in the *Journal of Consumer Research*, *Journal of Marketing Research*, *Marketing Science*, *International Journal of Research in Marketing* and the *Journal of the Market Research Society*, as well as journals in economics, psychology and management science; the IPA's '*Advertising Works*' series; and other academic and trade journals in marketing. Abstracts of 456 selected sources are given in Broadbent (1994). Vakratsas and Ambler (1995) give an integrated review of the literature.

The NFA report almost entirely ignores this literature. It also ignores much of the core literature on the direct issue of advertising and children, including the evidence on children's understanding of advertising. Where it does cite the literature, it tends to do so selectively and, in some cases, misleadingly (Sharp, 1994).

For instance, the section of the report titled 'The effect of advertising on children's food preferences' (p. 16) starts with the phrase 'There is much evidence to show that children are highly influenced by advertising in general'. This conclusion is attributed to reference 111 which cites three sources: Mintel (1991), Euromonitor (1988) and Greenberg, Fazal and Wober (1986). However, the only evidence from the Mintel report was the qualitative statement that 'respondents [i.e. mothers with children] had little doubt that advertisements had a strong effect on their children' as shown by the children's ability to recall TV commercials. But this fails to distinguish (a) recall from influence (b) brands from categories. We have already noted that the fact that an advertisement is recalled – or even liked – does not necessarily mean that it influences consumption, especially at the category level. These same respondents also said that advertising was only the **seventh** most important influence on their own general choice when shopping; just 14% said they were influenced by advertising (Mintel, 1991; Sharp, 1994).

Greenberg, Fazal and Wober (1986) concluded that 'young viewers [4–13 years] are quite favourable toward adverts on television', are 'more knowledgeable of adverts they like than those they dislike', and 'recognise that adverts exist to sell them something'. They were also 'sceptical' about the extent to which advertisements are truthful. Advertisements did 'induce young viewers to request product purchases from their parents', in many cases successfully. These requests were presumably for brands not categories (this was not one of the research questions). If these young viewers were 'highly influenced' by advertising, this was only in a very limited sense and with their own knowledge.

Neither Mintel (because its evidence is so shaky) nor Greenberg, Fazal and Wober (because their conclusions are more balanced than the NFA report's) provide 'much evidence that chilren are highly influenced by advertising in general'. I have not seen the third source (Euromonitor, 1988).

Other citations are subject to similar flaws in interpretation or method.

Perhaps, more importantly, there is also a considerable literature which the NFA report has ignored and which directly contradicts its claims. The literature covers many countries and many different product areas where linkages have been examined between the behaviour of children and product consumption, e.g:

- 'It is generally recognised that children's food habits are shaped by the nutritional ideals and dietary styles of the immediate family. The family functions as a micro culture by filtering from the larger culture the nutritional ideals and diet patterns that it deems appropriate for the nutritional health and gustatory satisfaction of its members. Within this context, the young child cultivates the food tastes and diet patterns of the family and learns strategies of food selections, procurement, preparation, distribution, and exchange.' (Jerome and Frese, 1979).
- 'The group observed that research generally supports the notion that advertising plays a relatively minor role in affecting attitudes, beliefs and behaviors, compared to interpersonal influences, such as family and reference groups.' (Ward, 1990).
- 'However, when asked about influence, especially influence on behaviour, the traditional authority figures are easily the most important accredited source. Parents' and families' opinions and actions are overwhelmingly the most influential across the board.' (The Human Factor, 1993).

8.5.2 Children's influence on food purchases

The NFA report (p. 19) states that 'Market research companies recognise that children are a major influence on household food purchases[121] and that their influence is increasing[122]. Ref. 121 (Taylor Nelson, 1991) does not say that children are a 'major' influence, nor does ref. 122 (Leatherhead Food Research Association, 1991) say that their influence is increasing (Sharp, 1994). Furthermore, the latter says that 'in the main, the influence appears to be a moderate one'.

The NFA report refers to Galst and White's (1976) small-scale study of purchase-influencing attempts (PIAs) during grocery shopping in the USA. But it does not point out that (a) it is not clear how many of the PIAs were for brands or particular items, as opposed to categories; (b) the 15 PIAs referred to included meats, fruits, vegetables and dairy products, as well as cereals and candy (in fact, as many PIAs were made for vegetables as for candy); and (c) as Galst and White themselves admit, the (fairly low) observed correlation between TV viewing and the number of PIAs does not demonstrate causality. Both may reflect underlying causal variables such as social class and parenting style (Carlson and Grossbart, 1988). (Since so many of the PIAs were for non-advertised products, this is almost certainly the case although it cannot be tested with Galst and White's data.)

Yet again, despite the report's claim on page 20, Rossiter (1981) neither said nor suggested that, as children get older, 'commercials make them more acquisitive'. He specifically concluded that 'TV advertising has no lasting cognitive effects'. He also said that:

'Arguments against TV advertising to children based on charges of generalized deception and youthful gullibility are simply not supported by the evidence' (Rossiter, 1981, p. 232).

The report (p. 19) notes that advertisers 'are not just promoting products, they are advertising brands' and quotes a mother in the Leatherhead study: 'I buy Sainsbury's Cola . . . but what they really want is brand names . . . They really want Pepsi.'

However, the NFA report fails to see the implications. What advertisers try to do (in this case unsuccessfully) is encourage people to choose their brand. This has little or no effect on total category sales. Children have always liked, say, sweets. What seems to have changed in the last 30–40 years is not children's category preferences (except in response to new categories) but their ability to influence adults to take these preferences into account. But this influence, which is still quite limited, reflects other social changes (McNeal, 1992).

8.5.3 Children's understanding of advertising

Other key claims in this section of the NFA report centre around the ability of children to understand advertising messages. Once again, the 'evidence' is selective and highly questionable.

For example, the section entitled 'Children's understanding of advertising' (p. 21) includes the following paragraph:

'Children's ability to understand advertising varies enormously according to age, with young children predictably the most impressionable. Zuckerman and Gianinno[138] found that three-quarters of four-year-olds were unable to differentiate between programmes and adverts. Neither could over a third of seven-year-olds or one in five ten-year-olds.'

The original source (Zuckerman and Gianinno, 1981) reported research on a small sample of middle class surburban American children. The authors do not say that the children sampled 'were unable to differentiate between programmes and adverts' but that when questioned they were 'unable to specify [i.e. articulate] a difference'. Verbal questioning (with verbal responses) was only one approach in the study: others were identification and recognition, paired comparisons and picture sorting. The authors were at pains to point out [p. 93]:

' . . . children's understanding of commercials was better than their verbal responses alone would allow us to infer. This disparity between verbal and non-verbal responses was particularly strong for the four-year-olds. The findings of the present study were consistent with the cognitive development literature which suggests that young children are often unable to express what they know; . . .

[references given]. Consequently, it is potentially misleading for researchers to infer from verbal responses alone what children do or do not perceive and understand about television programs and commercials' (underline added).

Indeed, this measurement issue was the main focus of Zuckerman and Giannino's study which was designed to determine 'whether the way questions are asked will change conclusions about how well children understand the differences between television programmes and television commercials' (p. 84).

Two paragraphs later, the NFA report says of the same study: 'even by the age of 10 only 15 per cent of the children knew that commercials intend to sell products'. Closer examination of the original paper reveals that, when asked 'what is a commercial?', answers from three ten-year-olds (15% of the sample) were indeed classified under 'Intent to sell (e.g. "They show you things you can buy")'. But answers from another twelve (or four times as many, or 60% in the NFA report's terms) were categorized as 'Show or describe something (e.g. "An advertisement", with no further clarification, or "Like when they show you a toy or something")'. This suggests that by the age of 10 as many as 75% know that commercials intend to sell products, even using only these verbal measures.

What makes this example worth exploring at some length is the fact that the claimed proportions of children 'unable to differentiate' are repeated in the NFA report's summary, where the context clearly refers to the UK, without any qualification at all.

Research that the NFA report has ignored includes, for instance, Gaines and Esserman's (1981) finding that

'children as young as 4 years old can easily distinguish between television programs and commercials, can recognize the selling intent of commercials and are even sceptical about commercials.'

In a more recent review, Goldstein (1992) concluded that

'The age at which children are found to understand television and advertising depends upon the method [of research] used. Research relying upon verbal responses puts the age of understanding higher (4 to 7 years old) than research using non verbal measures (as young as 3 years old)'.

Children too young to understand that the aim of advertising is to sell the advertised products (or strictly, brands) are also too young to have much influence on food purchases. Again, such influence as they do have in response to advertising is at the brand, not the category, level.

To my knowledge, the most thorough recent review of the literature on children's understanding of advertising, and advertising's influence on children, is by Young (1990). Young refers to experiments where

'Children are assigned to different groups where they watch particular commercials under different conditions. Afterwards they are given a choice from a range of snacks which include the advertised brand. Under these somewhat artificial conditions researchers often, but not always, find that the advertising of brand X makes the subsequent choice of brand X by the child more likely.

There are various reasons why this should occur, but with this research we are really no closer to finding an answer to the question of whether advertising snacks on television causes children to eat more snacks. This methodology has high internal validity but low external or ecological validity' (1990, p. 11).

Young (1990, p. 10) also refers to the evidence from Atkin (1975) and Galst and White (1976), already cited, and from Stoneman and Brody (1982) that 'children ask for food products advertised on television and that parents often acquiesce', but concludes that 'television advertising is a "weak force" when compared with other influences, such as family and friends. Although the existence of special audiences, where advertising plays a more influential role, cannot be ruled out, these susceptible groups have yet to be discovered and defined'.

8.6 Discussion

The production and consumption of most manufactured products involve some negative side-effects. Even with something as innocent as a tennis racquet one can pull a muscle, make the neighbours envious or disturb them by playing at first light, or even beat one's children. With a motor car, a kitchen knife, or a brick, one can kill.

But common sense suggests that we should allow people freely to buy, sell and advertise tennis racquets, cars, kitchen knives and bricks. We try to educate people to understand how to use these products sensibly. In the case of cars we also insist on them passing a driving test and wearing a seat belt. For really dangerous products like guns and powerful drugs, we try to regulate the market closely – and if we fail to do so, as with guns in the USA, the results can be disastrous.

The foods and drinks advertised on children's television are not dangerous in this sense. Of course, it would be unhealthy to live on nothing but, say, chocolate, just as it would be unhealthy to live on nothing but lentils. No single food dominates a varied and balanced diet.

In contrast, the anti-advertisers' case seems to be based on an assumption that the foods and drinks advertised to children are in themselves 'unhealthy'. Manufacturers and many independent nutritionists disagree. I have touched on this question, and on the evidence on children's health and diet, in section 8.3. Children's health, including dental health, has improved greatly over the last 30–40 years, with two almost opposite exceptions – obesity and nutrient deficiencies.

The increased incidence of obesity has not been caused by people eating more, but by reduced exercise among both children and adults. Energy intake has decreased, but not enough.

At the other extreme, a significant proportion of children are showing poor growth and low bodyweight resulting from insufficient intake of energy (not of micronutrients). The main cause is poverty. Another, especially among teenage girls, is deliberate 'dieting' in order to be slim, sometimes with serious or even tragic consequences.

Neither of these real-world health problems seems consistent with the idea that advertising is a major influence on nutrition (although one might argue that, without advertising, snacking would have increased less and energy intake decreased more, reducing obesity). In sections 8.4 and 8.5 we directly addressed this central question – the influence of advertising on children's diet. I believe the evidence is clear that advertising is not a major influence, relative to innate taste preferences and the strong influence of family and friends. The evidence is not clearcut, in the sense that it is not clear whether advertising's influence should be described as 'minor', 'minimal', or negligible'. But what is clear is that it is not 'major'. An important part of this conclusion is the distinction between brands and categories: advertising is mostly about brands, nutrition is mostly about categories.

8.6.1   The role of the food industry

'One of the unfortunate consequences of the NFA's aggressively confrontational stance is that it has become more difficult to have an intelligent, balanced debate on this important subject' (Feldwick, 1995, p. 67).

Given the lobbyists' enormously exaggerated idea of the power of advertising (not only as a 'negative' influence on diet but also as a potentially 'positive' one), their selection and distortion of evidence, their confused logic (e.g. they have still failed to address the distinction between brands and categories), their naive and damaging proposals, and the seductiveness of these proposals to politicians – whether well-meaning, cynical, or both – the natural response of the food industry is to adopt a similarly extreme position.

The temptation is to deny that advertising – and, even, the marketing mix as a whole (product, packaging, pricing, promotion, distribution) – has any influence at all on children's diet. In particular, the tendency in this context (as in the debates about tobacco and alcohol) is to argue that advertising is only about brand share and that there are no category effects. In section 8.4, I argued that this is not the case, listing six contexts in which advertising can lead to increased category sales, especially if the category is new and/or narrowly defined.

Similarly, I believe that the rest of the marketing mix can influence children's diet – although much less than the lobbyists suppose, because consumer choice is usually between quite close substitutes (e.g. different sweet snacks) rather than between, say, confectionery and fruit. Nevertheless, I agree with Feldwick (1995, p. 68) that 'Among the food industry's responsibilities is the responsibility to make it easy for people to eat healthily' – and without paying a premium unless this is based on genuinely higher costs. Manufacturers should have an explicit strategy to address the issue of children's diet – if only for reasons of enlightened self-interest, to pre-empt regulation.

Within this strategy, advertising should have a role to play in the introduction of new, healthier products – as it has over the last 30–40 years – as well as healthy eating habits.

8.6.2   The role of government

In my view, governments (and the EU) should refuse to let the anti-advertising lobbyists sidetrack them into treating advertising as a central issue in this debate, just because it is an easy target for legislation. There is a role for careful regulation of advertising execution – not encouraging 'grazing', not making misleading health claims – but restrictions on children's food and drink advertising *per se* will achieve nothing, reduce the amount of children's programming, make it harder to work in partnership with the food industry and, above all, miss the point.

The main influences on children's diet – inherent taste preferences, parents, other adults, peers – are hard for regulators to influence. Moreover, the social changes that have led to such marked changes in eating habits, including the much greater influence that children themselves now exert on what they eat, are not open to legislation. But, in addition to regulating advertising execution, positive steps can be taken to improve children's health and diet.

These steps include the availability of school meals: for instance, these are far more nutritious in French schools than in Britain (Blythman, 1995). Another area for improvement is school sports, whose decline has contributed to children's lack of fitness, the habit of taking no exercise, poorer health and obesity. Improving these areas will involve much work and thought, but (unlike a ban on advertising) will actually contribute to better health and diet among children.

Finally, governments can achieve gradual long-term changes through better consumer awareness programmes and by working with industry to influence the whole marketing mix. This lies beyond the scope of this chapter.

This chapter is a revised version of 'Children, Advertising, and Nutrition' (Barwise, 1994), a commentary on the 1993 National Food Alliance Report (Dibb, 1993).

References

Aaker, D.A. (1991), *Managing Brand Equity*, Free Press, New York.

Adamson, A., Rugg-Gunn, A.J., Butler, T., Appleton, D.R. and Hackett, A. (1992), Nutritional intake, height and weight of 11–12-year-old Northumbrian children in 1990 compared with information obtained in 1980, *Br. J. Nutrition*, **51**, 67–75.

Addy, D.P. (1986), Happiness is: iron. *Br. Medical J.*, **292**, 969–70.

Advertising Association (1993), *Review of Arguments and Data Relating to the NFA Report 'Children: Advertisers' Dream, Nutrition Nightmare'?*, Advertising Association, London.

Atkin, C.K. (1975), Effects of Television Advertising on Children: Second Year Experimental Evidence. In *Children and Faces of Television: Teaching, Violence, Selling*, Palmer, E.L. and Dorr, A. (Eds), Academic Press, New York, pp. 287–305.

Baker, C. (1993), *Advertising Works 7*, NTC, Henley-on-Thames.

Barwise, P. (1994), *Children, Advertising, and Nutrition: A Commentary on the National Food Alliance report 'Children: Advertisers' Dream, Nutrition Nightmare?'*. Advertising Association, London.

Bjorndal, T., Salvares, K.G. and Andreason, J.H. (1992), The Demand for Salmon in France: The Effects of Marketing and Structural Change. *Appl. Economics*, **24**(9), 1027–34.

Blythman, J. (1995), Aujourd'hui, we learn how to eat properly. *The Independent*, 11 March.

Broadbent, S. (Ed.) (1994), *456 Views of How Advertising Works*. Centre for Marketing and Communication, London Business School.

BTHA (1990), *Children and Advertising: A Survey*, British Toy and Hobby Manufacturers' Association.

Carlson, L. and Grossbart, S. (1988), Parental Style and Consumer Socialization of Children. *J. Consumer Res.*, June, 77–94.

Castell, A. (1988), 'The rattle of a stick in a swill bin'. Post Graduate Diploma in Health Education. (NFA report ref. 93).

Chamberlin, E.H. (1965), *The Theory of Monopolistic Competition*, Harvard University Press, Cambridge MA.

Cole-Hamilton, I., Dibb, S. and O'Rourke, J. (1991), *Fact Sheet: School Meals*, Child Poverty Action Group, London.

Consumer's Association (1992), School Dinners: Are they worth having? *Which?* September.

Davidson, J.H. (1972), *Offensive Marketing*, Cassell (Pelican, 1975).

Dibb, S.E. (1993), *Children: Advertisers' Dream, Nutrition Nightmare?* National Food Alliance, London.

Dickson, P.R. and Ginter, J.L. (1987), Market Segmentation, Product Differentiation, and Marketing Strategy. *J. Marketing*, **51** (April), 1–10.

Douglas, M. (1984), *Food in the Social Order: Studies of Food and Festivities in Three American Communities*, Russell Sage Foundation, New York.

Dowell, T.B. and Evans, D.J. (1989), The Dental Caries Experience of 5-Year-Old Children in Great Britain. *Community Dental Health*, **6**, 271–9.

Durnin, J.V.G.A., Lonergan, M.E., Good, J. and Ewan, A. (1974), A cross-sectional and anthropometric study, with an interval of 7 years, on 611 young adolescent schoolchildren. *Br. J. Nutrition*, **32**, 169–79.

ECDU (1987), *Children's diets in disadvantaged areas*, Early Childhood Development Unit, University of Bristol.

Ehrenberg, A.S.C. (1974), Repetitive Advertising and the Consumer. *J. Advertising Res.*, **14**(2), 25–34.

Ehrenberg, A.S.C., Goodhardt, G.J. and Barwise, T.P. (1990), Double Jeopardy Revisited. *J. Marketing*, **54**(3), 82–91.

Engel, J.E., Blackwell, R.D. and Miniard, P.W. (1990), *Consumer Behavior* (6th edn), Dryden Press, Chicago.

Escalona, S.K. (1945), Feeding disturbances in very young children. *Amer. J. Orthopsychiatry*, **15**, 76–80.

Esserman, J.F. (1981), Introduction: New Research vs Old Assumptions. In *Television Advertising and Children: Issues, Research and Findings*, Esserman, J.F. (Ed.), Child Research Service, New York.

Euromonitor (1988), *Children as consumers: marketing and market for the under 16s*, Euromonitor Publications.

Feldwick, P. (1995), Food, Advertising and the Nation's Diet: What Are the Real Issues? *Business Strategy Rev.*, Spring, 61–9.

Fox, K.R. (1993), 'Do children take enough exercise?' Paper presented at 'Children in Focus: a National Dairy Council Conference', London, 28 October.

Gaines, L. and Esserman, J.F. (1981), A Quantitative Study of Young Children's Comprehension of Television Programs and Commercials. In *Television Advertising and Children: Issues, Research and Findings*. Esserman, J.F. (Ed.) Child Research Service, New York.

Galst, J.P. and White, M.A. (1976), The unhealthy persuader: the reinforcing value of television and children's purchase influencing attempts at the supermarket. *Child Development*, **47**, 1089–96.

Gardner, B.B. and Levy, S.J. (1955), The product and the brand. *Harvard Business Rev.*, March–April.

Gardner Merchant (1990), *School meals survey*, Survey report.

Girling, R. (1995), The Crunch. *The Times Magazine*, 4 March, 26–9.

Goldberg M.E. (1990), A Quasi-Experiment Assessing the Effectiveness of TV Advertising Directed to Children. *J. Marketing Res.*, **27** (November), 445–54.

Golder, P.N. and Tellis, G.J. (1993), Pioneer Advantage: Marketing Logic or Marketing Legend? *J. Marketing Res.*, **30**(2), 158–70.

Goldstein, J.H. (1992), *Television Advertising and Children: A Review of Research*, Toy Manufacturers of Europe, Brussels.

Gray, J. (1991), *Advertising Bans: Administrative Decisions or Matters of Principal?* Social Affairs Unit, London.

Green, R.D., Carman, H.F. and McManus, K. (1991), Some Empirical Methods of Estimating Effects in Demand Systems: An Application to Dried Fruits. *Western J. Agricultural Econ.*, **16**(1), 63–71.

Greenberg, B.S., Fazal, S., and Wober, M. (1986), *Children's Views on Advertising*, Independent Broadcasting Authority, London.

Guber, S.S. and Berry, J. (1993), *Marketing to and Through Kids*, McGraw-Hill, New York.

Health Education Authority (1990a), *Sugars in the Diet*, Briefing paper, Health Education Authority, London.

Health Education Authority (1990b), *Young adults' health and lifestyle: diet*, Health Education Authority, London.

Henry, H. (1984), Does advertising affect total market size? *Admap*, **20**(11), 524–32.

HMSO (1960), *The Health of School Children*, Report of the Chief Medical Officer of the Ministry of Education 1955–59, Her Majesty's Stationery Office, London.

Human Factor, The (1993), 'Advertising Association research on the influences on children's food preferences', report by The Human Factor.

Iceland (1990), 'Attitudes to children and food'. Survey report.

Jacoby, J. and Chestnut, R.W. (1978), *Brand Loyalty: Measurement and Management*, Wiley, New York.

Jensen, H.H. and Schroeter, J.R. (1992), Television Advertising and Beef Demand: An Econometric Analysis of 'Split-Cable' Household Panel Scanner Data. *Canadian J. Agricultural Econ.*, **40**, 271–94.

Jerome, N.W. and Frese, D.J. (1979), What are the relative contributions of family and television to a child's food preference? In *Proceedings of the Sixth Annual Telecommunications Policy Research Conference*, Dordick, H.S. (Ed.). Lexington Books, Lexington M.A.

King, S. (1973), *Developing New Brands*, Pitman, London.

Kotler, P. (1994), *Marketing Management: Analysis, Planning, Implementation, and Control* (8th edn), Prentice-Hall.

Kunkel, D. and Roberts, D. (1991), Young Minds and Marketplace Values: Issues in Children's Television Advertising. *J. Social Issues*, **47**(1), 57–72.

Lambin, J.J. (1976), *Advertising, Competition and Market Conduct in Oligopoly over Time*, North-Holland, Oxford.

Lancaster, K. (1979), *Variety, Equity, and Efficiency*, Columbia University Press, New York.

Lean, M.E.J., James, W.P.T. and Garthwaite, P.H. (1989), Obesity without overeating. In *Obesity in Europe 88*, Bjorntorp, P. and Rossner, S. (Eds), John Libbey, 281–6.

Leatherhead Food Research Association (1991), *Children's Eating Habits. An in-depth study of the attitudes and behaviour of children aged 6–11*, LFRA.

Lipsey, R.G. (1983), *An Introduction to Positive Economics* (6th edn), Weidenfeld and Nicolson.

Lunn, T. (1986), Segmenting and constructing markets. In *Consumer Market Research Handbook* (3rd edn) Worcester, R. and Downham, J. (Eds), McGraw-Hill.

Mars, G. and Mars, V. (1993), *Food, Culture and History* (Volume 1), The London Food Seminar, London.

McDonald, C. (1992), *How Advertising Works: A Review of Current Thinking*, The Advertising Association/NTC, London.

McNeal, J.U. (1992), *Kids as Customers*, Lexington Books, New York.

Mercer, D. (1992), *Marketing*, Blackwell, Oxford.

Mintel (1991), *Children: The Influencing Factor*, Special Report, Mintel.

Nelson, M. (1991) Nutrition and the schoolchild: Food vitamins and IQ. *Proc. Nutrition Soc.*, **50**, 29–35.

Nelson, M., White, J. and Rhodes, C. (1992), Haemoglobin, ferritin and iron intakes in British children aged 12–14 years: a preliminary investigation. *Br. J. Nutrition* (NFA ref. 23, in press).

Nuttall, N. (1992), Video trains children to eat detested healthy food. *The Times*, 25 August.

Porter, M.E. (1976), *Interbrand Choice, Strategy, and Bilateral Market Power*, Harvard University Press, Cambridge MA.

Portman Group (1993), 'Britain's Young Adults' (18–30) Attitudes to Drink-Driving'. Survey by Audience Selection Ltd, London for: The Portman Group.

Prentice, A.M. and Jebb, S.A. (1995), Obesity in Britain; Gluttony or Sloth? *Br. Medical J.*, **311** (August), 437–9.

Ries, A. and Trout, J. (1985), *Positioning The Battle for Your Mind*, McGraw-Hill, New York.

Ritchey, N. and Olsen, C. (1983), Relationships between family variables and children's preference for consumption of sweet foods. *Ecology of Food and Nutrition*, **13**, 257–66.

Robertson, T.S. (1971), *Innovative Behavior and Communication*, Holt, Rinehart & Winston, New York.

Rogers, E.M. (1983), *Diffusion of Innovations*, (3rd edn), Free Press, New York.

Rona, R.J. and Chinn, S. (1984), The National Survey of Health and Growth: Nutritional Surveillance of Primary School Children from 1972–81 with special reference to unemployment and social class. *Annals of Human Biol.*, **11**, 17–28.

Rossiter, J. (1981), Research on Television Advertising's General Impact on Children: American and Australian Findings. In *Television Advertising and Children: Issues, Research and Findings*, Esserman, J.F. (Ed.), Child Research Service, New York.

Rousseau, N. (1984), 'Bites and pieces'. PhD thesis, University of Edinburgh, UK.

Sharp, C. (1994), *An Analysis of the References Used in 'Children: Advertisers' Dream, Nutrition Nightmare*? Advertising Association, London.

Simon, J.L. (1970), The Effect of Advertising upon the Propensity to Consume. In *Issues in the Economics of Advertising*, University of Illinois Press.

Stoneman, Z. and Brody, G.H. (1982), The Indirect Impact of Child-Oriented Advertisements on Mother–Child Interactions. *J. Appl. Developmental Psychology*, **2**, 369–76.

Taylor Nelson (1991), *The Young Ones, Children's Food and Drink Consumption: The Marketing Opportunities*. Family Food Panel Special Report.

Thomas, J. (1991), Food choices and preferences of schoolchildren. *Proc. Nutrition Soc.*, **50**, 49–57.

Vakratsas, D. and Ambler, T. (1995), *The Effects of Advertising*, Centre for Marketing Working Paper, London Business School.

Venkateswaran, M. and Kinnucan, H.W. (1990), Evaluating Fluid Milk Advertising in Ontario: The Importance of Functional Form. *Canadian J. Agricultural Econ.*, **38**(3), 471–88.

Walters, C. and Ambler, T. (1993), *Advertising Effect on Mature Market Size*, PAN' AGRA Working Paper, London Business School.

Ward, S. (1990), The effects of tobacco advertising on adolescent smoking initiation and smoking maintenance: overview of a group seminar. *Int. J. Advertising*, **9**, 85–91.

Yasin, J. (1995), The Effects of Advertising on Fast-Moving Consumer Goods Markets. *Int. J. Advertising*, **14**, 133–47.

Young. B. (1990), The Role of Advertising in the Life of the Child: A Review of some Empirical Findings. *Int. J. Advertising*, **9**, 1–14.

Zidack, W., Kinnucan, H. and Hatch, U. (1992), Wholesale-Level and Farm-Level Impacts of Generic Advertising: The Case of Catfish. *Appl. Econ.*, **24**(9), 959–68.

Zuckerman, P. and Gianinno, L. (1981), Measuring Children's Responses to Television Advertising. In *Television Advertising and Children: Issues, Research and Findings*, Esserman, J.F. (Ed.), Child Research Service, New York.

9 Children's views on food and nutrition: a pan-European study

JEAN-PIERRE PROPONNET, CHAIRMAN OF THE
EUROPEAN FOOD INFORMATION COUNCIL
(EUFIC)

9.1 Chairman's introduction

The first, essential step in product innovation and marketing of children's food and drinks is an accurate understanding of their nutritional knowledge and perceptions. Much inaccurate, even biased 'misinformation' has been promoted by those who profess to represent the best interests of children and yet whose opinions are based more on personal 'hobbyhorses' than on any reliable, quantitative market research. For example, some experts argue that children's diet today is made up primarily of fast foods, carbonated soft drinks, confectionery and non-dairy ice cream! Others argue that children's knowledge and perceptions of nutritional food and drink values is low – or simply wrong. Yet others argue that TV advertising has a major adverse effect.

The truth, thankfully, is very different. A major pan-European survey has just been completed which addressed this very issue. Entitled 'Children's views on food and nutrition', it highlights a most encouraging level of common sense which is being reflected in the diets of many children. They demonstrated a lot more knowledge about what is good for them – and what is not so good – than many might have expected. Furthermore, their overall eating habits reflect this knowledge. Far from being 'junk food addicts', children today understand the importance of nutrition and the value of a balanced diet. This does not mean, however, that they don't admit to enjoying what is not so good for them – far from it! But it does mean they fully understand the need for moderation in their eating habits. This level of common sense is cause for celebration.

Founded in late 1993 at the initiative of major European food and drinks companies, the brief for the European Food Information Council (EUFIC) is to listen and respond to the questions and concerns of European consumers about the foods they eat. It is recognized that variety – both in food and in information about food – can lead to confusion about such important issues as nutrition, the safety and wholesomeness of food, the impact of food processing, the relative characteristics of various foods and, most important of all, the role of the consumer in ensuring balanced family diets as well as safe, healthy and enjoyable meals.

EUFIC undertook this survey in order to help assess the knowledge levels, perceptions and actions of young people on nutrition and other food and drink-related topics. Since diet is certainly amongst the most important determinants of health, how children understand nutrition today – and how they act on this understanding – will have an important impact on their health tomorrow. This research-based knowledge, rather than any mis-guided opinion, is a prerequisite for the development of a successful European-wide nutrition education programme. The findings, too, are of great value to those employed in the product development and marketing of successful food and drinks for the increasingly important and ever-growing children's market.

The survey was conducted on behalf of EUFIC in the four largest European countries: France, Germany, Italy and the United Kingdom between December 1994 and May 1995, through the internationally-respected Children's Research Unit (CRU). CRU and its chairman, Glen Smith, have more than 20 years experience in researching and reporting on key social, ethical and marketing issues amongst children and teenagers on behalf of governments, industry associations and councils and for many of the leading multi-national corporations.

For the EUFIC survey, personal interviews were conducted amongst a total of 1600 children, aged between 8–15-years-old, with 400 interviews being undertaken in each of the four countries. Since children across this age span are known to hold significantly different levels of understanding and articulation, research was conducted and reviewed in three age groups: 8–10-year-olds, 11–12-year-olds and 13–15-year-olds. A standard question-naire was used for all three age groups and multiple responses to some questions were encouraged. In each household, parental permission was obtained before the child interview was conducted, but the parent/ guardian was not present.

Previous research in the food/drinks arena, conducted by CRU, had shown that children do not believe any food or drink to be bad for them. Rather, they see food and drinks as having various degrees of goodness. Based upon this experience, which was confirmed by the pilot work undertaken for this survey, a three-point nutritional rating scale developed by CRU was used in order to elicit responses to a variety of issues in which their opinion on the goodness or otherwise of food and drinks was being sought. Children were asked to base their opinion on whether the particular food or drink was thought by them to be:

- 'good for you'
- 'OK for you'
- 'not so good for you'

Minor adjustments to the choice of words used in Italy for the three-point scale arose as a result of the pilot work undertaken there.

In each country, children were asked to nominate one meal they had eaten in the previous 24 hours, to list the food and drinks served at that meal as well as the key decision-maker. This provided not only a valuable 'snapshot' of the menus across Europe but also highlighted the differences in eating preferences, reflecting the differences in food culture. It was found that the menus in France and Italy had many similarities (influenced, no doubt, by the central role of the matriarch) but that these preferences differed from those found in Germany and UK where children exerted a greater influence on some menus – especially the evening meal in Germany and the lunchtime meal in the UK.

Separately, children were asked to rate a portfolio of food and drinks which included those listed. Not surprisingly, children identified as being 'good for you' many of the staple foods, such as fruit and vegetables, milk and milk-based products, bread, pasta and cereals, fish, meat, chicken and eggs. Equally, children rated foods less commonly eaten at mealtimes, such as processed food and treats, as being 'not so good for you'. Mum still knows best!

What was even more encouraging, however, was the recognition by children of the need for a balanced diet. 82% agreed that 'foods like sweets and ice cream are OK to eat, but not all the time' whilst 79% agreed 'it is best to eat small amounts of different foods, rather than a lot of the same food'. Whilst there can be no doubt as to the enjoyment, derived by many children right across the age spectrum, from eating fast foods and chocolates, these were recognized as being 'not so good for you'. Only 11% agreed that 'chocolate is OK to eat every day' whilst a surprisingly low 8% agreed that 'fast foods are OK to eat every day'.

Children demonstrated great confidence about their nutritional knowledge. Almost 2 out of every 3 children claimed to know 'a lot' or 'some things' about nutrition. Over 90% agreed it was important that children learn about nutrition. The credit for achieving this was given to the family (though, less so in UK), whilst the school/school teacher also played a central role (more so in UK). Inspite of being singled out by 'experts' as a prime cause of 'the problem', neither television nor advertising was nominated as influential in providing nutritional guidance by more than 1 in 5 children in this multiple-answer scenario.

This nutritional confidence was endorsed by responses to the importance of a list of prompted nutrients. Vitamins, protein and calcium were each nominated (multiple answers) by 4 in 5 children as needed by the body 'to stay healthy' whilst those thought to be of much less importance included starch, salt and fat. Just 7% nominated pills!

For the food industry, perhaps, the most disappointing levels of knowledge and perception arose for answers to questions about processed foods. Processed foods are recognized as very much a part of their everyday lives, yet children expressed far more uncertainty about their nutritional role. Less than 25% agreed that 'ready-made meals are just as good for you as home-made meals', whilst only a similar percentage agreed that 'tinned fruit and vegetables are just as good for you as fresh fruit and vegetables'. Food safety, too, proved to be a concern with 69% agreeing that 'fresh foods are safer than frozen or tinned foods'. This opinion was reflected in the belief by 29% that 'foods containing E-numbers are bad for you' – a statement on which a further 58% of children could make no comment. Indeed, the level of 'don't knows' was higher for questions about processed foods than other foods. To sum it all up, only 42% of children were able to agree that 'the food we eat today is safer than ever before'. These findings confirm the belief by some that there is still much educational work to be undertaken to improve the understanding of the nutritional value and safety of processed foods. EUFIC sees itself as being at the forefront of encouraging this educational initiative.

Yet, whilst some uncertainties were expressed about the safety of 'the food we eat today', understanding of food hygiene was encouragingly high. Response levels to a series of statements showed that >80% agreed with 4 different aspects of everyday food hygiene. Furthermore, 61% of children correctly identified 'harmful bacteria in food' as the cause of food poisoning whilst, in a separate question, 49% recognized Salmonella as 'a bacteria which, when found in food, can make you ill'. This knowledge was extended to a proper understanding of the meaning of 'use-by' dates on short shelf life products. 77% of children (with more in France and Italy) believed that 'use-by dates tell you the food must be eaten before that date'.

The EUFIC survey, the detail of which follows in this chapter, provides one of the first and most accurate 'windows' into that children across Europe understand and think about the nutritional values of food and drinks. The findings are most encouraging. Children have shown that they do have a good level of correctly based nutritional knowledge; they do understand which foods are good for them and which are not so good; they appreciate the need for balance and moderation in their diet; they know how to 'mix' staple foods with self-treats and they respect the importance of learning more about nutrition. Clearly, the healthy eating messages are being listened to and acted upon correctly by children.

They do, however, demonstrate less certainty about the 'values' of processed foods. Whilst it is possible to delve deeply into specific attitudes to specific food/drinks – of prime interest to the target audience for this book – the new product development manager and brand marketing manager – the major benefit from the EUFIC survey lies in providing invaluable guidance on broad principles which can help direct future thinking on nutritional education. This, in turn, needs to be tailored to satisfy differing EU cultural realities. These initiatives, in which nutrition information institutes, consumer organizations, government agencies, health professionals, teachers and the news media have a vital role to play, can help lead to a more healthy and better informed future for the children of tomorrow.

9.2   Background and research objectives

This pan-European study concerning children's views on food and nutrition was commissioned to obtain a basic understanding of the perceptions of European children on food- and drink-related issues. The study was based on a questionnaire designed to provide information on:

● current eating and drinking patterns;
● children's views on nutrition and health;
● where children learn about food and nutrition;
● how children view food safety.

9.3   Methodology

Research was focused on four countries: France, Germany, Italy and the UK. These countries – which have the largest populations of all EU member states – were selected in order to study similarities and differences in dietary habits across Europe.

Interviews were carried out with a total of 1600 children aged 8 to 15 years, with 400 interviews taking place in each of the four countries covered. Since children in this age bracket are known to show significant differences in understanding and the ability to express themselves, research was carried out on three age groups: 8–10-year-olds; 11–12-year-olds; and 13–15-year-olds.

A standard questionnaire was used in all four countries and across the three age groups (see Appendix A); and multiple responses were allowed for all questions marked with an asterisk (*) in the questionnaire. In each household contacted, parental permission was obtained to interview one child. Field work was carried out in December 1994, January 1995 and May 1995.

Responses in this report were not broken down by age, sex and social class because it was found that the differences in responses given across these segments were not overly significant, and that general trends emerged right across these groups.

9.4 Executive summary

9.4.1 Current eating and drinking patterns

Children were asked to give details of one of their meals during the 24 hours preceding the interview. Typical menus for breakfast, lunch, and dinner in these four countries show that interesting similarities exist between diets in France and Italy, as between those of Germany and the UK.

Breakfasts in France and Italy consist mostly of bakery products; likewise for Germany. In the UK, breakfast cereals top the ranking of the most frequently consumed breakfast foods, and they are in second place in Germany. When it comes to breakfast drinks, French children favour hot chocolate, Italians and Germans prefer milk, and British children drink tea the most frequently.

Almost one in five (19%) children in Italy does not eat anything for breakfast, a percentage which is much higher than in the other countries; in the UK it is 9%, in France, 5%, and just 1% in Germany.

At lunch, vegetables and salad are quite popular across all four countries, with over half of the children surveyed eating them. French and Italian children consume fruit much more frequently than children in Germany and the UK. French children eat yoghurt much more frequently than children elsewhere. Amongst carbohydrate-rich foods, pasta is popular in France and Italy, while potatoes are favoured in Germany and the UK. British children eat french fries much more often than their counterparts in the other countries. Sandwiches are particularly popular in the UK, where almost one-quarter of children eat them for lunch, compared with less than 6% in the other three countries. When it comes to protein-rich foods, French and Italians eat meat, chicken, fish, and cheese more often than children in the other two countries.

Water is the most popular lunchtime drink in France and Italy, with colas and lemonades being favoured in Germany and the UK.

The dinner menus of schoolchildren include a diverse range of foods. Amongst carbohydrate-rich foods, pasta is favoured in France and Italy, while German children prefer bread, and British children eat potatoes, especially french fries, more often. Of protein-rich foods, children in France, Germany, and the UK eat such foods more frequently than children in Italy; French and Italian children consume fish more frequently than the others; and German children eat cheese more often.

Children in France and Italy favour water as a drink with dinner, whilst children in Germany and the UK favoured a greater variety of drinks.

Children claim to have a significant influence on the choice of menu at breakfast. Across the four countries studied, 70% claim to make their own decisions on what to eat during the week, even though 48% eat this meal with their family. For the evening meal – both during the week and at weekends – only 28% claim to decide for themselves what to eat. School routines influence the decision-making of children at lunch time.

Nearly half of children surveyed in France, Germany and Italy eat their weekday breakfast with one or more parents. In the UK, 36% share this meal with parents. At weekends, around two-thirds of children in France and Italy eat breakfast with parents, compared with less than half of UK children and over four-fifths (82%) of German children.

During the school term, two-fifths (41%) of children in France, 56% in Germany, 77% in Italy and only 4% in the UK eat lunch with at least one parent. At the weekend, over 90% of children in France, Germany and Italy share lunch with a parent, compared with 58% in the UK.

Across all four countries studied, the vast majority of children eat the main evening meal with one or more parents. In France, 96% of children have their evening meal with the family both during the week and at weekends; in Germany, 88% eat as a family during the week, a figure which rises to 92% at the weekend; in Italy, 96% eat with parents on weekdays and 95% at the weekend; and in the UK, 87% of children eat with parents during the week, which rises to 91% at weekends.

9.4.2   Children's views on nutrition and health

Overall, children are pretty well-versed in understanding nutrition and recognizing the importance of a balanced diet for good health. The majority of children agree that 'Milk is good for strong bones' (84% agree); 'Foods like sweets and ice cream are OK to eat, but not all the time' (82%); 'To stay healthy, you should eat less fat' (82%); 'The food you eat affects your health as you grow up' (81%); 'Exercise is just as important as the foods you eat for staying healthy' (79%); 'It is important to eat foods like whole grain bread and cereals' (79%); and 'It is best to eat small amounts of different foods, rather than a lot of the same food' (74%). Only 11% agree that 'Chocolate is OK to eat every day', and 8% agree that 'Fast foods are OK to eat every day.'

Across the four countries, on average, children rate the importance of nutrients to health as follows: vitamins (94% of children agree that the body needs them in order to stay healthy), protein (82%), calcium (78%), minerals (70%), fibre (61%), calories (43%), sugar (42%), starch (38%), salt (34%), fat (33%) and pills (7%).

When asked what foods are 'Good for you', the most frequent answer is fruits (85%), followed by vegetables (82%), water (73%) and bread (72%). The four-country 'Not so good for you' ranking is led by beer (78%), wine (74%), sweets (69%) and cider (66%). Minor differences in these rankings were recorded in each country.

When asked to compare the nutritional value of fresh foods and processed foods, 61% disagree with the statement, 'Ready-made meals are just as good for you as home-made meals' and 60% disagree that 'Tinned fruit and vegetables are as good for you as fresh fruit and vegetables'. Almost one in six is unable to reply. Some 69% of children believe that 'Fresh foods are safer than tinned or frozen foods.'

9.4.3 Learning about food and nutrition

Most children understand that the word 'nutrition' is related to the health and dietary benefits of foods. However, close to one-third of children in France (32%) and the UK (34%) were unable to express any opinion on the word's meaning. This figure was much lower in Germany (6%) and Italy (3%).

Across all four countries, an average of 92% of children believe that it is important to learn about nutrition. Only 7% of children claim to 'know a lot' about nutrition. A further 56% 'know some things', 30% 'don't know much' and 7% 'don't know anything' about nutrition.

Overall, the family is perceived by 67% of children as being their most important source of information on nutrition. The child's school (41%) and school teachers (34%) are also perceived as playing a central role. Television programmes are a source of information for 17% of children. Children less often learn about nutrition from advertising, magazines and friends.

Half of all children agree that cooking classes at school would be the best way to learn about nutrition. Some 43% would prefer information packs for use at school, while 31% would like to learn more from television programmes.

9.4.4 Food safety and hygiene

Overall, children appear to understand the basic rules of food hygiene. The majority of children agree with the statements: 'You should always wash fruit and vegetables before eating them' (96% agree); 'It is important to keep the kitchen clean' (95%); 'Food can go "off" from being kept at the wrong temperature' (86%); and 'You should always cover food before putting it back in the fridge' (83%).

Some 77% of children are aware that 'use-by' dates on food mean that 'You need to eat the food before that date', whilst 8% do not understand this information.

When asked whether 'Foods containing E numbers are bad for you', 58% of children claim they do not know. 42% are in agreement with the statement 'The food we eat today is safer than ever before'; 31% do not know whether this is the case. 53% disagree with the statement, 'It is safer

to drink milk straight from the cow than the milk one buys in the store';
26% claim they do not know whether this is true.

Concerning food poisoning, 61% of children believe that it is caused by
'Harmful bacteria in food', whilst 28% attribute it to 'The pesticides that
are used on crops'. Some 49% are aware that Salmonella food poisoning is
caused by harmful bacteria in food, while 41% do not know what
Salmonella is.

9.5 Comparisons between countries: current eating and drinking patterns

9.5.1 What children eat and drink

> | Question: | I want to talk to you about food and nutrition, but, first of
> all, can you tell me what you had to eat and drink for breakfast today/lunch
> today/evening meal yesterday?

Each child was asked to give details of one meal eaten during the 24 hours
before the interview. He or she was able to select from 'breakfast today/
lunch today/evening meal yesterday'; interviewers kept a check that
overall, a balance of coverage between the different meals was obtained in
each country.

(a) Breakfast. Bakery products are the most frequently eaten breakfast
foods in France, Italy and Germany, while cereals dominate the menu of
many UK children (57%), although toast is also popular (26%) there. In
none of the four countries surveyed are yoghurt or fruit frequent breakfast
choices. Tea is a frequent breakfast drink for British (26%) and German
(21%) children; but consumed much less often in the other countries.
Italian children drink coffee much more frequently (17%) than do children
in the other three countries. Milk, either hot or cold, is the most universal
of the breakfast drinks. 48% of French children drink hot chocolate with
their breakfast.

In Italy, a significant percentage of children (19%) do not eat anything
for breakfast. This compares to 9% in the UK, 5% in France and just 1%
in Germany.

(b) Lunch. Vegetables and salad are very popular for lunch in each
country – on average, 55% of children eat them for lunch. Potatoes are
also eaten by many in Germany (28%) and especially the UK (41%), and
less in France (25%) and Italy (11%). In the UK, 30% have french fries at
lunch while almost half of Italian children (46%) eat pasta. These two
figures are much higher than in other countries.

Of protein-rich foods, meat is the most popular lunchtime choice. 56% of French children eat meat for lunch, which means that they consume it more often than children in other countries. Chicken and fish are not very popular lunchtime choices in any country.

Relatively few children in all countries have soup for lunch; the average is only 3%. Almost a quarter (24%) of UK children eat sandwiches, which compares to under 6% for France and Italy and none for Germany. Italy and France have a much higher consumption rate for fruit (30% and 24% respectively), than the UK at 5% and Germany at just 2%. 16% of French children eat yoghurt for lunch, while only 4% of British children do, and 0% of German and Italian children have it for lunch.

The vast majority of children in France (74%) and Italy (90%) drink water; just 21% in Germany and 15% in the UK do so. On average, a number of children choose colas (14%) and lemonades (11%). These two choices are especially prominent in the UK (each at 22%).

(d) Dinner. Meat retains its popularity at dinner time in France (44%), Germany (45%) and the UK (44%), but fewer children have it in Italy (24%). 37% of German children have sausages for dinner, and 56% of British children eat potatoes. These two figures are considerably higher than comparable percentages in the other countries. French fries are a popular (34%) dinner choice for children in the UK.

Some country-specific dinner trends are evident. Pasta appears on 24% of dinner tables in Italy, compared to only 19% in France, 6% in the UK, and 5% in Germany. Almost half (46%) of German children eat bread at dinner, whereas the four-country average is 19%. 30% of French children have yoghurt at dinner, compared to the four-country average of just 9%.

Vegetables and salad are most popular in the UK (68%) followed by France (43%), Italy (39%) and Germany (27%). More children in the UK (12%) and Italy (10%) have chicken than in Germany (3%) and France (2%), while fish is more popular in Italy (16%) and France (13%) than in the UK (8%) and Germany (4%). 23% of German children have cheese at dinner, the highest percentage for this product of the four countries (France 17%, Italy 14% and the UK 4%). The popularity of soup at dinner is higher than at lunch, but is still low, with an average of just 7%.

Fruit consumption is again much higher in Italy and France, with 26% of Italian children eating it and 14% of the French. This compares to just 7% of UK children and only 3% of German children.

71% of children in France have water. This compares to 81% in Italy, 14% in the UK and 9% in Germany; Tea is quite popular in Germany (15%) and the UK (14%), as is cola (17% for Germany, 12% for the UK). In both Germany and the UK, a wide variety of drinks appear at dinner, including lemonade, milk and fruit juice. In the UK, this list is complemented by squashes of assorted flavours.

9.5.2 Meal choices – who makes the decisions?

| Question: | *Who usually decides what you eat for:*

- *breakfast during the week;*
- *breakfast at the weekend;*
- *lunch during the week;*
- *lunch at the weekend;*
- *dinner/supper during the week;*
- *dinner/supper at the weekend.*

Choice: Father/I do/Mother/School/Other

One meal was selected for each child. Multiple responses account for the fact that figures do not always add up to 100%.

Children claim to have significant influence on the choice of menu at breakfast. Across the four countries studied, 70% claim to make their own decisions on what to eat on weekdays (Table 9.5.1). At weekends, an average of 69% of children claim to choose their breakfast menu (Table 9.5.2). For lunch and dinner, the chief decision-maker is the mother, who makes choices on behalf of the majority of children both during the week and at weekends (Tables 9.5.1 and 9.5.2). Responses from this survey suggest that fathers play little or no part in selecting food for their children.

Table 9.5.1 'Who usually decides what you eat?' (during the week)

%	Breakfast		Lunch		Dinner	
	I do	Mother	I do	Mother	I do	Mother
F	75	23	8	41	12	86
G	65	37	23	69	63	47
I	60	40	15	80	19	84
UK	78	20	49	29	19	79
Average	70	30	24	55	28	74

Table 9.5.2 'Who usually decides what you eat?' (during the weekend)

%	Breakfast		Lunch		Dinner	
	I do	Mother	I do	Mother	I do	Mother
F	74	23	13	83	14	84
G	63	39	17	84	60	50
I	60	39	18	81	20	81
UK	77	21	45	55	18	77
Average	69	31	23	76	28	73

9.5.3 Meals eaten with the family

| Question: | Which meals do you usually eat with your family?

Dinner is the meal eaten most frequently with the family (by family, we mean with at least one parent present), both during the week and at weekends (Table 9.5.3). Lunch, particularly at weekends, is also often eaten with one or both parents. During the week, differing schooling patterns in the four countries account for variations in eating habits at lunchtime, with 41–77% of children in France, Germany and Italy eating with their families, compared with a figure of only 4% for the UK. Breakfast is eaten with the family most frequently at weekends, although almost half of children across the four countries eat with one or both parents during the week.

Table 9.5.3 'Which meals do you eat with your family?'

%	Breakfast		Lunch		Dinner	
	W/D	W/E	W/D	W/E	W/D	W/E
F	49	64	41	94	96	96
G	48	82	56	94	88	92
I	58	68	77	96	96	95
UK	36	37	4	58	87	91
Average	48	65	45	86	92	94

W/D = weekday, W/E = weekend

9.6 Comparisons between countries: children's views on nutrition and health

9.6.1 Perceptions of food processing and its effects on nutrition and health

| Question: | I would now like to read you a number of statements. Please tell me if you agree or disagree with these statements or if you don't know:

Statements:

- 'Ready-made meals are just as good for you as home-made meals'
- 'Tinned fruit and vegetables are just as good for you as fresh fruit and vegetables'
- 'Fresh foods are safer than frozen or tinned foods'
- 'It is safer to drink milk straight from the cow than the milk one buys in the store'
- 'The food we eat today is safer than ever before'
- 'Foods containing E numbers are bad for you'

Across the four countries, the majority of children (61%) disagree with the statement 'Ready-made meals are just as good for you as home-made meals' (Table 9.6.1). Just under one-quarter agree with the statement and about one-sixth don't know.

Table 9.6.1 Statement: 'Ready-made meals are just as good for you as home-made meals'

%	Agree	Disagree	Don't know
F	25	66	9
G	18	52	30
I	13	81	6
UK	34	43	24
Average	23	61	17

Most children (60%) disagree with the statement that 'Tinned fruit and vegetables are just as good for you as fresh fruit and vegetables' (Table 9.6.2). Just under one-quarter agree with it and around one-sixth do not know.

Table 9.6.2 Statement: 'Tinned fruit and vegetables are just as good for you as fresh fruit and vegetables'

%	Agree	Disagree	Don't know
F	25	68	8
G	20	50	30
I	15	78	7
UK	33	43	23
Average	23	60	17

The majority (69%) of children believe that 'Fresh foods are safer than tinned or frozen foods' (Table 9.6.3). Around one-seventh disagree with this statement and just under one-fifth don't know.

Table 9.6.3 Statement: 'Fresh foods are safer than frozen or tinned foods'

%	Agree	Disagree	Don't know
F	74	17	9
G	55	14	31
I	84	10	6
UK	63	14	24
Average	69	14	18

Over one-half (53%) of children disagree with the statement 'It is safer to drink milk straight from the cow than the milk one buys in the store', whilst just under one-quarter agree with the statement and slightly more than one-quarter don't know (Table 9.6.4).

Table 9.6.4 Statement: 'It is safer to drink milk straight from the cow than the milk one buys in the store'

%	Agree	Disagree	Don't know
F	27	50	23
G	24	32	44
I	27	59	14
UK	8	69	23
Average	22	53	26

Many children (42%) agree that 'The food we eat today is safer than ever before', whilst just over one-quarter disagree with the statement, and slightly less than one-third do not know (Table 9.6.5).

Table 9.6.5 Statement: 'The food we eat today is safer than ever before'

%	Agree	Disagree	Don't know
F	61	19	20
G	21	29	50
I	34	46	20
UK	51	14	35
Average	42	27	31

The majority (58%) of children do not know whether 'Foods containing E numbers are bad for you', whilst 29% agree with the statement and 13% disagree with it (Table 9.6.6).

Table 9.6.6 Statement: 'Foods containing E numbers are bad for you'

%	Agree	Disagree	Don't know
F	19	17	65
G	22	8	69
I	44	9	47
UK	29	18	52
Average	29	13	58

9.6.2   How children rate the nutritional quality of foods and drinks

| Question: | *I have a list of foods and drinks here and I want you to show me which ones you think are 'Good for you', 'Okay for you' or 'Not so good for you'.*

A three-point Nutritional Rating Scale developed by CRU was used to evaluate children's knowledge of nutrition. Since previous research indicates that children do not believe any food or drink to be 'bad for you', this Nutritional Rating Scale allows interviewees to rate the perceived nutritional value of individual food and drink products (from a prompted list) by indicating whether they believe a food to be:

- 'Good for you';
- 'OK for you';
- 'Not so good for you'.

In Italy, the three-point scale was amended in order to further clarify the association between health and nutrition. Here, pilot research found that the questionnaire was most clearly understood by interviewees when the three-point scale used the following three options: 'Very good for your health'; 'Good for your health'; and 'Not so good for your health'.

Classifying foods, especially when involving children, sometimes poses difficulties. In this survey, as a result of pilot studies, potatoes are not included as vegetables, french fries are not included as potatoes, nor is chicken considered as meat; in children's minds, these foods were distinct and separate.

Across the four countries, children most frequently rated 'Good for you' foods and drinks as being fruits, followed by vegetables, water and milk (Tables 9.6.7, 9.6.8). The 'Not so good for you' ranking is led by beer, followed by wine, sweets and cider (Tables 9.6.9, 9.6.10).

9.6.3   Attitudes towards health and nutrition

| Question: | *I would now like to read you a number of statements. Can you tell me if you agree or disagree with these statements, or if you don't know:*

Statements:

- *'It is best to eat small amounts of different foods rather than a lot of the same food'*

Table **9.6.7** Foods and drinks which children believe to be 'Good for you'

%	Fruit	Vegetables	Water	Milk
F	94	92	96	89
G	81	76	51	69
I	71	65	60	54
UK	94	94	84	75
Average	85	82	73	72

Table 9.6.8 Four-country averages: Foods and drinks which children believe to be 'Good for you'

Rank	Food/Drink	%*
1	Fruit	85
2	Vegetables	82
3	Water	73
4	Milk	72
5	Bread	60
6	Fish	59
7	Chicken	58
8	Cheese	56
9	Pasta	52
10	Soup	51
11	Meat	50
12	Breakfast cereal	49
13	Eggs	43
14	Pizza	27
15	Butter	25
16	French fries	19
16	Margarine	19
18	Biscuits	18
19	Ice cream	15
19	Burgers	15
21	Cakes	14
21	Ketchup	14
23	Sugar	13
23	Chocolates	13
23	Fizzy drinks	13
26	Crisps	12
27	Sweets	7
28	Cider	6
29	Beer	4
29	Wine	4

*Percentage of children who rate it as 'Good for you'

Table 9.6.9 Foods and drinks which children believe to be 'Not so good for you'

%	Beer	Wine	Sweets	Cider
F	78	71	65	57
G	75	82	37	79
I	85	78	88	57
UK	72	63	84	69
Average	78	74	69	66

- 'To stay healthy, you should eat less fat'
- 'Milk is good for strong bones'
- 'Foods like sweets and ice cream are OK to eat, but not all the time'
- 'It is important to eat foods like whole grain bread and cereals'
- 'Fast foods are OK to eat every day'

Table 9.6.10 Four-country averages: Foods and drinks which children believe to be 'Not so good for you'

Rank	Food/Drink	%*
1	Beer	78
2	Wine	74
3	Sweets	69
4	Cider	66
5	Fizzy drinks	59
6	Crisps	58
7	Chocolate	56
8	Sugar	49
9	French fries	47
9	Ketchup	47
11	Burgers	46
12	Cakes	44
13	Ice cream	43
14	Butter	38
14	Margarine	38
16	Biscuits	36
17	Pizza	25
18	Eggs	13
19	Breakfast cereal	12
20	Pasta	8
20	Meat	8
20	Fish	8
20	Cheese	8
24	Soup	7
25	Water	6
26	Chicken	5
27	Bread	4
28	Vegetables	3
28	Milk	3
30	Fruit	1

*Percentage of children who rate it as 'Not so good for you'

- 'The food you eat affects your health as you grow up'
- 'Exercise is just as important as the food you eat for staying healthy'
- 'Chocolate is OK to eat every day'

Almost three-quarters (74%) of children agree that 'It is best to eat small amounts of different foods, rather than a lot of the same food' (Table 9.6.11). Some 11% disagree with this statement, and 15% don't know.

Table 9.6.11 Statement: 'It is best to eat small amounts of different foods rather than a lot of the same food'

%	Agree	Disagree	Don't know
F	82	11	8
G	67	6	27
I	79	13	8
UK	67	14	18
Average	74	11	15

Over four-fifths (82%) of children agree that 'To stay healthy, you should eat less fat' (Table 9.6.12). Some 7% disagree with this statement, and 11% don't know.

Table 9.6.12 Statement: 'To stay healthy, you should eat less fat'

%	Agree	Disagree	Don't know
F	84	8	9
G	76	4	19
I	85	9	6
UK	83	8	10
Average	82	7	11

The majority of children (84%) agree that 'Milk is good for strong bones', with only 4% disagreeing and 12% saying they 'don't know' (Table 9.6.13).

Table 9.6.13 Statement: 'Milk is good for strong bones'

%	Agree	Disagree	Don't know
F	84	5	11
G	77	4	19
I	88	4	8
UK	86	4	10
Average	84	4	12

Most children (82%) agree that 'Foods like sweets and ice cream are OK to eat, but not all the time' (Table 9.6.14). Just over one-tenth disagree and 6% don't know.

Table 9.6.14 Statement: 'Foods like sweets and ice cream are OK to eat, but not all the time'

%	Agree	Disagree	Don't know
F	84	13	3
G	82	8	10
I	81	17	2
UK	80	11	8
Average	82	12	6

Most children (79%) agree that 'It is important to eat foods like whole grain bread and cereals', with one-tenth disagreeing, and 12% saying they 'don't know' (Table 9.6.15).

Table 9.6.15 Statement: 'It is important to eat foods like whole grain bread and cereals'

%	Agree	Disagree	Don't know
F	82	9	9
G	78	7	15
I	69	18	13
UK	85	4	11
Average	79	10	12

Four-fifths of children disagree with the statement that 'Fast foods are OK to eat every day', and less than one-tenth agree with it (Table 9.6.16). Some 13% don't know.

Table 9.6.16 Statement: 'Fast foods are OK to eat every day'

%	Agree	Disagree	Don't know
F	7	91	3
G	6	69	24
I	5	85	10
UK	14	73	13
Average	8	80	13

Over four-fifths (81%) of children agree that 'The food you eat affects your health as you grow up', with just under one-tenth disagreeing and one-tenth saying they 'don't know' (Table 9.6.17).

Table 9.6.17 Statement: 'The food you eat affects your health as you grow up'

%	Agree	Disagree	Don't know
F	84	6	11
G	67	19	14
I	94	4	2
UK	80	7	13
Average	81	9	10

Almost four-fifths (79%) agree that 'Exercise is just as important as the food you eat for staying healthy', whilst one-tenth disagree with this statement, and the remainder don't know (Table 9.6.18).

Over four-fifths (82%) disagree with the statement 'Chocolate is OK to eat every day', and just over one-tenth agree with the statement, while 7% 'dont know' (Table 9.6.19).

Table 9.6.18 Statement: 'Exercise is just as important as the food you eat for staying healthy'

%	Agree	Disagree	Don't know
F	85	10	6
G	67	8	24
I	80	15	5
UK	85	5	10
Average	79	10	11

Table 9.6.19 Statement: 'Chocolate is OK to eat every day'

%	Agree	Disagree	Don't know
F	14	83	3
G	15	73	13
I	8	88	4
UK	7	84	9
Average	11	82	7

9.6.4   Understanding how nutrients relate to health

Question: *To stay healthy, which of these does your body need?*

Choice: Protein, fat, minerals, vitamins, salt, fibre, pills, calories, calcium, sugar, starch

Across the four countries, vitamins are most widely recognized as being important for health (94%), followed by protein (82%), calcium (78%), minerals (70%), fibre (61%), calories (43%), sugar (42%), starch (38%), salt (34%), fat (33%) and pills (7%) (Table 9.6.20).

Table 9.6.20 'To stay healthy, which of these does your body need?'

%	Vita-mins	Pro-tein	Calcium	Min-erals	Fibre	Cal-ories	Sugar	Starch	Salt	Fat	Pills
F	98	91	91	84	70	52	62	47	58	45	12
G	94	66	69	71	50	36	30	45	37	39	3
I	93	93	82	51	51	44	53	23	18	23	4
UK	91	78	71	75	73	39	24	35	22	23	7
Average	94	82	78	70	61	43	42	38	34	33	7

9.7 Comparisons between countries: learning about food and nutrition

9.7.1 Children's interpretations of the word 'nutrition'

| Question: | *What does the word 'nutrition' mean to you?*

Most children understand that the word 'nutrition' is related to the health and dietary benefits of foods. However, close to one-third of children in France (32%) and the UK (34%) are unable to express any opinion as to the word's meaning. This figure is much lower in Germany (6%) and Italy (3%).

9.7.2 What children think they know about nutrition

| Question: | *How much do you think you know/don't know about nutrition? Can you show me which of these statements applies to you the most?*

Choice:

- *I know a lot about nutrition.*
- *I know some things about nutrition.*
- *I don't know much about nutrition.*
- *I don't know anything about nutrition.*

In total, only 7% of children claim to 'know a lot' about nutrition. More than one-half (56%) 'know some things', 30% 'don't know much' and 7% 'don't know anything' about nutrition (Table 9.7.1).

Table 9.7.1 'How much do you know about nutrition?'

%	A lot	Some	Not much	Don't know anything
F	8	61	25	7
G	10	54	30	5
I	5	66	27	3
UK	5	44	37	14
Average	7	56	30	7

9.7.3 Attitudes towards learning

| Question: | *Do you think it is important that children learn about nutrition?*

Across the four countries, an average of 92% of children believe it important to learn about nutrition (Table 9.7.2). This ranges from 83% in

Table 9.7.2 'Do you think it is important that children learn about nutrition?'

%	Yes	No
F	94	6
G	83	17
I	97	3
UK	92	8
Average	92	8

Germany to 97% in Italy. An average of 8% of children do not believe it important to learn about nutrition.

9.7.4 Current sources of information on nutrition

| Question: | Where do you learn about nutrition? |

Choice: *Family/friends/schoolteacher/television/school/magazines/advertising*

A distinction was made between learning about nutrition at school and learning about it from a teacher because children perceive the two as being somewhat different; the child could actually learn from the teacher, or from educational materials such as books, leaflets, etc. available at school.

Overall, the family is perceived by 67% of children as being their most important source of information on nutrition (Table 9.7.3). The child's school (41%) and school teachers (34%) also play a central role. Television programmes are a source of information for 17% of children. Children in the UK said that they most frequently learn about nutrition from school. They also learn much less frequently about nutrition from their family, in contrast with children in other countries where the family is the primary information source.

Table 9.7.3 'Where do you learn about nutrition?'

%	Family	School	Teacher	Television	Advertising	Magazines	Friends
F	74	35	32	12	10	10	4
G	78	41	29	30	20	15	13
I	81	38	37	11	7	8	3
UK	35	49	36	16	9	7	2
Average	67	41	34	17	12	8	4

9.7.5   Preferred future methods of learning about nutrition

| Question: | How would you like to learn about nutrition?

Choice: Cooking classes at school/television programmes/magazines/special information packs for use at school

On average, 50% of children across the four countries would, in the future, like to obtain information on nutrition from cooking classes (Table 9.7.4). Information packs for use at school are favoured by 43% of children and TV programmes by just under one-third.

Table 9.7.4 'How would you like to learn about nutrition?'

%	Cooking classes	Information packs at school	TV programmes
F	42	38	28
G	52	23	31
I	49	68	41
UK	56	41	25
Average	50	43	31

9.8   Comparisons between countries: food safety and hygiene

9.8.1   Understanding of food hygiene

| Question: | I would now like to read you a number of statements. Please tell me if you agree or disagree with these statements or if you don't know.

Statements:

- *'It is important to keep the kitchen clean'*
- *'You should always cover food before putting it back in the fridge'*
- *'You should always wash fruit and vegetables before eating them'*
- *'Foods can go "off" from being kept at the wrong temperature'*
- *'You don't always need to wash your hands before you eat'*

The vast majority (95%) of children agree that 'It is important to keep the kitchen clean' (Table 9.8.1). The great majority (83%) also agree that 'You should always cover food before putting it back in the fridge' (Table 9.8.2).

Almost all children (96%) agree that 'You should always wash fruit and vegetables before eating them' (Table 9.8.3).

Most children (86%) agree that 'Food can go "off" from being kept at the wrong temperature' (Table 9.8.4).

Table 9.8.1 Statement: 'It is important to keep the kitchen clean'

%	Agree	Disagree	Don't know
F	95	3	3
G	91	2	7
I	99	1	1
UK	95	1	4
Average	95	2	4

Table 9.8.2 Statement: 'You should always cover food before putting it back in the fridge'

%	Agree	Disagree	Don't know
F	79	10	12
G	78	5	17
I	86	4	10
UK	88	3	9
Average	83	6	12

Table 9.8.3 Statement: 'You should always wash fruit and vegetables before eating them'

%	Agree	Disagree	Don't know
F	96	3	2
G	96	1	3
I	99	1	0
UK	91	1	7
Average	96	2	3

Table 9.8.4 Statement: 'Food can go "off" from being kept at the wrong temperature'

%	Agree	Disagree	Don't know
F	86	6	8
G	89	2	9
I	83	7	10
UK	85	3	12
Average	86	5	10

The majority of children (84%) disagree with the statement 'You don't always need to wash your hands before you eat', with only around one-seventh agreeing with it (Table 9.8.5).

Table 9.8.5 Statement: 'You don't always need to wash your hands before you eat'

%	Agree	Disagree	Don't know
F	16	83	1
G	16	79	5
I	8	92	0
UK	14	80	6
Average	14	84	3

9.8.2   Understanding of 'use-by' dates

| Question: | Which of these things do 'use-by' dates tell you?

Choice:

- *You need to eat the food before that date.*
- *It's OK to eat it two weeks after the date mentioned.*
- *It's the date by which stores need to sell that food.*
- *Don't know.*

Over three-quarters (77%) of children are aware that 'use-by' dates on food mean that 'You need to eat the food before that date', whilst 8% claim that they do not know what this means (Table 9.8.6).

Table 9.8.6 'Which of these things do "use-by" dates tell you?'

%	Eat before date	OK to eat 2 weeks after date	Store sell-by date	Don't know
F	75	4	18	5
G	67	8	22	9
I	96	0	2	2
UK	71	3	11	14
Average	77	4	13	8

9.8.3   Understanding of food poisoning

| Question: | How do you get food poisoning?

Choice:

- *From eating too much.*
- *From harmful bacteria in food.*

- *From the additives in food.*
- *From the pesticides that are used on crops.*
- *Don't know.*

Multiple responses were allowed.

Across the four countries studied, most (61%) children believe that food poisoning is caused by 'Harmful bacteria in food', whilst over one-quarter attribute it to 'The pesticides that are used on crops' (Table 9.8.7). One-seventh believe it to be due to food additives and one-sixth attribute it to eating too much.

Table 9.8.7 'How do you get food poisoning?'

%	Bacteria	Pesticides	Additives	Eating too much	Don't know
F	43	20	13	30	21
G	80	28	12	12	9
I	56	50	28	24	4
UK	65	15	2	2	17
Average	61	28	14	17	13

9.8.4   Understanding of Salmonella

| Question: | *What is Salmonella?*

Choice:

- *Bacteria, which, when found in food, make you ill.*
- *A type of pink salmon.*
- *A common childhood disease.*
- *Don't know.*

Just under one-half (49%) of all children agree that Salmonella is a form of harmful bacterium found in food, but this figure varies widely across the four countries studied, from 20% in France to 69% in Germany (Table 9.8.8). An average of slightly more than two-fifths of children (41%) don't know, although this percentage rises to 65% in France.

Table 9.8.8 'What is Salmonella?'

%	Bacteria found in food	Pink salmon	Child disease	Don't know
F	20	1	14	65
G	69	0	1	29
I	56	6	14	24
UK	50	3	3	44
Average	49	3	8	41

9.9   France: current eating and drinking patterns

9.9.1   What children eat and drink

| Question: | *I want to talk to you about food and nutrition, but, first of all, can you tell me what you had to eat and drink for breakfast today/lunch today/evening meal yesterday?*

Each child was allowed to give details of one meal eaten during the 24 hours before the interview. He or she was able to select from breakfast today/lunch today/evening meal yesterday; interviewers kept a check that overall, a balance of coverage between the different meals was obtained in each country.

At breakfast, bakery products are the most popular choice, with 70% of French children eating them. 19% eat cereals. Some 60% of children have a hot drink for breakfast, mainly hot chocolate, which is preferred by just under half of French children (48%). Just over one-quarter (27%) drink cold milk, and 13% prefer orange juice. 5% leave home without eating breakfast.

At lunch, meat is a popular food, with 56% of French children eating it, followed by poultry (14%) and fish (13%). 69% of children eat vegetables and salad, and one-quarter have potatoes. Some 17% of children eat pasta for lunch, with 15% eating French fries. 17% have cheese and 24% have fruit (as in Italy, this figure is much higher than in the other two countries). The majority of children (74%) drink water with their lunch, with 11% opting for cola.

At dinner, some 44% of French children eat meat, 13% eat fish and 4% eat poultry. Vegetables and salad are eaten by 43% of children, with 14% eating potatoes. Just under one-fifth (19%) eat pasta, 14% eat bakery products and 8% eat sandwiches. Some 17% have cheese and 14% have fruit. 71% of children drink water at dinner.

One of the most striking differences between French children and their counterparts from the other three countries is the high level of yoghurt consumption at lunch (16%) and at dinner (30%). This compares to the low four-country averages of 5% for lunch and 9% for dinner.

9.9.2   Meal choices – who makes the decisions?

| Question: | *Who usually decides what you eat for:*

- *breakfast during the week;*
- *breakfast at the weekend;*
- *lunch during the week;*

- *lunch at the weekend;*
- *dinner/supper during the week;*
- *dinner/supper at the weekend.*

Choice: Father/I do/Mother/School/Other

For the above questions, each child was asked to comment on one meal. Some multiple responses account for the fact that figures do not always add up to 100%.

Like children in the other three countries studied, French children claim to have the greatest influence on the choice of menu at breakfast, with close to three-quarters selecting their meal both during the week and at weekends (Table 9.9.1). This is slightly higher than the four-country average figure of 70% during the week and 69% at weekends. Also in common with other countries (except Germany), the main decision-maker at dinner is the mother, who makes choices for most children, especially at weekends. Decision-making at lunch during the week depends upon schooling patterns; in France, 49% of children have school lunches.

Table 9.9.1 'Who usually decides what you eat?'

%	Breakfast		Lunch		Dinner	
	I do	Mother	I do	Mother	I do	Mother
During the week						
France	75	23	8	41	12	86
Four-country average	70	30	24	55	28	74
During the weekend						
France	74	23	13	83	14	84
Four-country average	69	31	23	76	28	73

9.9.3 Meals eaten with the family

| Question: | *Which meals do you usually eat with your family?*

In common with the pattern seen in the other countries surveyed, French children most frequently eat dinner with the family (by this, we mean at least one parent present), both during the week and at weekends (Table 9.9.2). Lunch, especially at weekends, is also often eaten with one or more parent. On weekdays, around two-fifths of children eat lunch with their families, whilst at weekends, the vast majority do. Breakfast is eaten with the family by just under half (49%) of French children during the week. At weekends, 64% of French children eat breakfast with the family.

Table 9.9.2 'Which meals do you eat with your family?'

%	Breakfast		Lunch		Dinner	
	W/D	W/E	W/D	W/E	W/D	W/E
France	49	64	41	94	96	96
Four-country average	48	65	45	86	92	94

W/D = weekday, W/E = weekend

9.10   France: children's views on nutrition and health

9.10.1   Perceptions of food processing and its effects on nutrition and health

| Question: | *I would now like to read you a number of statements. Please tell me if you agree or disagree with these statements or if you don't know:*

Statements:

- *'Ready-made meals are just as good for you as home-made meals'*
- *'Tinned fruit and vegetables are just as good for you as fresh fruit and vegetables'*
- *'Fresh foods are safer than frozen or tinned foods'*
- *'It is safer to drink milk straight from the cow than the milk one buys in the store'*
- *'The food we eat today is safer than ever before'*
- *'Foods containing E numbers are bad for you'*

In France, two-thirds of children disagree with the statement 'Ready-made meals are just as good for you as home-made meals', with one-quarter agreeing with it (Table 9.10.1). These figures are close to the four-country average.

Table 9.10.1 Statement: 'Ready-made meals are just as good for you as home-made meals'

%	Agree	Disagree	Don't know
France	25	66	9
Four-country average	23	61	17

In France, as in the other countries examined, most children (68%) disagree with the statement that 'Tinned fruit and vegetables are just as good for you as fresh fruit and vegetables' (Table 9.10.2).

Table 9.10.2 Statement: 'Tinned fruit and vegetables are just as good for you as fresh fruit and vegetables'

%	Agree	Disagree	Don't know
France	25	68	8
Four-country average	23	60	17

Almost three-quarters of French children (74%) believe that 'Fresh foods are safer than tinned or frozen foods' (Table 9.10.3). Overall, some 17% disagree and just under one-tenth don't know.

Table 9.10.3 Statement: 'Fresh foods are safer than frozen or tinned foods'

%	Agree	Disagree	Don't know
France	74	17	9
Four-country average	69	14	18

One-half of French children disagree with the statement 'It is safer to drink milk straight from the cow than the milk one buys in the store' (Table 9.10.4). Overall, just over one-quarter agree with the statement and slightly less than one-quarter don't know.

Table 9.10.4 Statement: 'It is safer to drink milk straight from the cow than the milk one buys in the store'

%	Agree	Disagree	Don't know
France	27	50	23
Four-country average	22	53	26

Over three-fifths (61%) of French children agree that 'The food we eat today is safer than ever before' (Table 9.10.5), a figure which is substantially higher than the four-country average of just over two-fifths (42%). Almost one-fifth of French children disagree with it, whilst one-fifth don't know.

As in the other countries, the majority of French children (65%) don't

Table 9.10.5 Statement: 'The food we eat today is safer than ever before'

%	Agree	Disagree	Don't know
France	61	19	20
Four-country average	42	27	31

Table 9.10.6 Statement: 'Foods containing E numbers are bad for you'

%	Agree	Disagree	Don't know
France	19	17	65
Four-country average	29	13	58

know whether 'Foods containing E numbers are bad for you' (Table 9.10.6), a figure which is rather higher than the four-country average of 58%. Overall, just under one-fifth agree with the statement, whilst 17% disagree.

9.10.2   How children rate the nutritional quality of foods and drinks

| Question: | *I have a list of foods and drinks here and I want you to show me which ones you think are 'Good for you', 'Okay for you' or 'Not so good for you'.*

A three-point Nutritional Rating Scale developed by CRU was used to evaluate children's knowledge of nutrition. Since previous research indicates that children do not believe any food or drink to be 'bad for you', this Nutritional Rating Scale allows interviewees to rate the perceived nutritional value of individual food and drink products (from a prompted list) by indicating whether they believe a food to be:

- 'Good for you';
- 'OK for you';
- 'Not so good for you'.

Classifying foods, especially when involving children, sometimes poses difficulties. In this survey, as a result of pilot studies, potatoes are not included as vegetables, French fries are not included as potatoes, nor is chicken considered as meat; in children's minds, these foods were distinct and separate.

In France, water leads the 'Good for you' rankings, followed by fruit, yoghurt and vegetables (Table 9.10.7). This differs slightly from the four-country average ranking, which is led by fruits, followed by vegetables, water and milk.

French children's 'Not so good for you' rankings are led by beer, wine, sweets and burgers (Table 9.10.8), differing only a little from the four-country average ranking, which is led by beer, followed by wine, sweets and cider.

Table 9.10.7 France: Foods and drinks which children believe to be 'Good for you'

Rank		Percentage of children who rate it as 'Good for you'	Four-country average (%)
1	Water	96	73
2	Fruit	94	85
3	Yoghurt	93	–
4	Vegetables	92	82
5	Milk	89	72
5	Soup	89	51
7	Cheese	84	56
8	Chicken	80	58
9	Fish	77	59
10	Bread	75	60
10	Breakfast cereal	75	49
12	Meat	72	50
13	Pasta	62	52
14	Eggs	59	43
15	French fries	33	19
16	Pizza	32	27
17	Butter	30	25
18	Biscuits	23	18
19	Sugar	22	13
20	Margarine	20	19
20	Cakes	20	14
22	Ice cream	18	15
23	Chocolate	16	13
24	Fizzy drinks	15	13
24	Burgers	15	15
24	Ketchup	15	14
27	Crisps	13	12
27	Cider	13	6
29	Sweets	8	7
30	Wine	7	4
31	Beer	4	4

9.10.3   Attitudes towards health and nutrition

Question: *I would now like to read you a number of statements. Can you tell me if you agree or disagree with these statements, or if you don't know?*

Statements:

- *'It is best to eat small amounts of different foods rather than a lot of the same food'*
- *'To stay healthy, you should eat less fat'*
- *'Milk is good for strong bones'*
- *'Foods like sweets and ice cream are OK to eat, but not all the time'*
- *'It is important to eat foods like whole grain bread and cereals'*

Table 9.10.8 France: Foods and drinks which children believe to be 'Not so good for you'

Rank		Percentage of children who rate it as 'Not so good for you'	Four-country average (%)
1	Beer	78	78
2	Wine	71	74
3	Sweets	65	69
4	Burgers	57	46
4	Cider	57	66
6	Crisps	55	58
7	Fizzy drinks	52	59
8	Ketchup	51	47
9	Ice cream	46	43
10	Chocolate	42	56
11	Margarine	41	38
12	Sugar	40	49
13	Cakes	33	44
14	Butter	32	38
15	Biscuits	31	36
16	French fries	27	47
17	Pizza	22	25
18	Pasta	11	8
19	Eggs	8	13
20	Breakfast cereal	5	12
21	Fish	4	8
22	Cheese	3	8
22	Soup	3	7
24	Bread	2	4
24	Milk	2	3
24	Chicken	2	5
27	Water	1	6
27	Vegetables	1	3
27	Fruit	1	1
27	Meat	1	8
27	Yoghurt	1	–

- *'Fast foods are OK to eat every day'*
- *'The food you eat affects your health as you grow up'*
- *'Exercise is just as important as the food you eat for staying healthy'*
- *'Chocolate is OK to eat every day'*

Following the pattern seen in the other countries studied, over four-fifths of French children (82%) agree that 'It is best to eat small amounts of different foods, rather than a lot of the same food (Table 9.10.9). Overall, some 11% of children disagree with this statement, and 8% don't know.

Table 9.10.9 Statement: 'It is best to eat small amounts of different foods rather than a lot of the same food'

%	Agree	Disagree	Don't know
France	82	11	8
Four-country average	74	11	15

In line with interviewees in the other countries surveyed, over four-fifths (84%) of French children agree that 'To stay healthy, you should eat less fat' (Table 9.10.10). Some 8% disagree with this statement, and 9% don't know.

Table 9.10.10 Statement: 'To stay healthy, you should eat less fat'

%	Agree	Disagree	Don't know
France	84	8	9
Four-country average	82	7	11

As in the other countries, most French children (84%) agree that 'Milk is good for strong bones', with only 5% disagreeing and 11% saying they 'don't know' (Table 9.10.11).

Table 9.10.11 Statement: 'Milk is good for strong bones'

%	Agree	Disagree	Don't know
France	84	5	11
Four-country average	84	4	12

Following the same trend seen in the other three countries, the majority of French children (84%) agree that 'Foods like sweets and ice cream are OK to eat, but not all the time' (Table 9.10.12). Some 13% disagree and 3% don't know.

Table 9.10.12 Statement: 'Foods like sweets and ice cream are OK to eat, but not all the time'

%	Agree	Disagree	Don't know
France	84	13	3
Four-country average	82	12	6

Over four-fifths of French children (82%) agree that 'It is important to eat foods like whole grain bread and cereals' (Table 9.10.13). 9% of children disagree with this statement, and 9% don't know. This follows the pattern seen in the other countries.

Table 9.10.13 Statement: 'It is important to eat foods like whole grain bread and cereals'

%	Agree	Disagree	Don't know
France	82	9	9
Four-country average	79	10	12

The vast majority (91%) of French children disagree with the statement that 'Fast foods are OK to eat every day', the highest figure seen in the four countries studied (Table 9.10.14). Some 7% agree with the statement and 3% don't know.

Table 9.10.14 Statement: 'Fast foods are OK to eat every day'

%	Agree	Disagree	Don't know
France	7	91	3
Four-country average	8	80	13

In line with the pattern seen in the other countries, over four-fifths (84%) of French children agree that 'The food you eat affects your health as you grow up', with 6% disagreeing and 11% saying they 'don't know' (Table 9.10.15).

Table 9.10.15 Statement: 'The food you eat affects your health as you grow up'

%	Agree	Disagree	Don't know
France	84	6	11
Four-country average	81	9	10

As in the other countries, the majority (85%) of French children agree that 'Exercise is just as important as the food you eat for staying healthy' (Table 9.10.16). One-tenth of children disagree with this statement, and 6% don't know.

Table 9.10.16 Statement: 'Exercise is just as important as the food you eat for staying healthy'

%	Agree	Disagree	Don't know
France	85	10	6
Four-country average	79	10	11

Over four-fifths (83%) of French children disagree with the statement 'Chocolate is OK to eat every day' (Table 9.10.17), with one-seventh agreeing with the statement, and 3% saying they 'don't know'. Similar responses were obtained in the other countries.

Table 9.10.17 Statement: 'Chocolate is OK to eat every day'

%	Agree	Disagree	Don't know
France	14	83	3
Four-country average	11	82	7

9.10.4   Understanding how nutrients relate to health

| Question: | To stay healthy, which of these does your body need? |

Choice: Protein, fat, minerals, vitamins, salt, fibre, pills, calories, calcium, sugar, starch

The nutrients most widely recognized as being important for health by French children were vitamins (98%), protein (91%), calcium (91%) and minerals (84%) (Table 9.10.18). These follow the same order as the four-country average figures, with the French percentages being consistently higher than average.

Table 9.10.18 Nutrients most widely recognized as being important for health

	France	Four-country average (%)
Vitamins	98	94
Calcium	91	78
Protein	91	82
Minerals	84	70
Fibre	70	61
Sugar	62	42
Salt	58	34
Calories	52	43
Starch	47	37
Fat	45	32

9.11   France: learning about food and nutrition

9.11.1   Children's interpretations of the word 'nutrition'

| Question: | What does the word 'nutrition' mean to you? |

A rather low number (4%) of French children understand that the word 'nutrition' connotes 'Good for Health/Healthy eating'; 13% say that the word 'Has to do with food'. Close to one-third of children in France (32%) were unable to express any opinion on the word's meaning. This compares to figures of 6% in Germany, 3% in Italy and 34% in the UK.

9.11.2   What children think they know about nutrition

| Question: | How much do you think you know/don't know about nutrition? Can you show me which of these statements applies to you the most?

Choice:

- *I know a lot about nutrition.*
- *I know some things about nutrition.*

- *I don't know much about nutrition.*
- *I don't know anything about nutrition.*

Claimed knowledge of nutrition among French children is in line with the four-country averages (Table 9.11.1), with 8% claiming to know 'a lot' (compared with the average of 7%). Some 61% know 'some things' (compared with the average of 56%). One-quarter claim to know 'not much' about nutrition (compared with the average of 30%). Some 7% – equal to the four-country average – said that they know nothing.

Table 9.11.1 'How much do you think you know/don't know about nutrition?'

%	A lot	Some things	Not much	Nothing
France	8	61	25	7
Four-country average	7	56	30	7

9.11.3   Attitudes towards learning

| Question: | *Do you think it is important that children learn about nutrition?*

94% of French children believe it important that they learn about nutrition, a figure slightly higher than the four-country average of 92% (Table 9.11.2). Only 7% do not believe it important to learn about nutrition.

Table 9.11.2 'Do you think it is important that children learn about nutrition?'

%	Yes	No
France	94	7
Four-country average	92	9

9.11.4   Current sources of information on nutrition

| Question: | *Where do you learn about nutrition?*

Choice: Family/friends/school teacher/television/school/magazines/advertising

A distinction was made between learning about nutrition at school and learning about it from a teacher because children perceive the two as being somewhat different; the child could actually learn from the teacher, or from educational materials such as books, leaflets, etc. available at school.

As in the other countries studied, the family is seen as the most important source of nutritional information by French children, with just under three-quarters (74%) saying that they learn about this topic at home (Table 9.11.3). This figure is slightly higher than the four-country average of 67%. The next most important source of information in France is the school (mentioned by over one-third of children), followed by the teacher (32%) and television (12%).

Table 9.11.3 'Where do you learn about nutrition?'

%	Family	School	Teacher	Television	Advertising	Magazines	Friends
France	74	35	32	12	10	10	4
Four-country average	67	41	34	17	12	8	4

9.11.5   Preferred future methods of learning about nutrition

| Question: | *How would you like to learn about nutrition?*

Choice: Cooking classes at school/television programmes/magazines/special information packs for use at school

In France, 42% of children believe that cooking classes at school would be the best way to learn more about nutrition in the future, compared with a four-country average of 50% (Table 9.11.4). Just under two-fifths (38%) would like to use information packs at school (compared with the average figure of 43%), and 28% believe that television should play a role, compared with the average figure of 31%.

Table 9.11.4 'How would you like to learn about nutrition?'

%	Cooking classes	Information packs/leaflets	TV programmes
France	42	38	28
Four-country average	50	43	31

9.12   France: food safety and hygiene

9.12.1   Understanding of food hygiene

| Question: | *I would now like to read you a number of statements. Please tell me if you agree or disagree with these statements or maybe you don't know.*

Statements:

- *'It is important to keep the kitchen clean'*
- *'You should always cover food before putting it back in the fridge'*
- *'You should always wash fruit and vegetables before eating them'*
- *'Foods can go "off" from being kept at the wrong temperature'*
- *'You don't always need to wash your hands before you eat'*

The great majority (95%) of French children agree that 'It is important to keep the kitchen clean'; this 95% is identical to the four-country average (Table 9.12.1).

Table 9.12.1 Statement: 'It is important to keep the kitchen clean'

%	Agree	Disagree	Don't know
France	95	3	3
Four-country average	95	2	4

79% of children in France agree that 'You should always cover food before putting it back in the fridge', a figure which is somewhat lower than the four-country average of 83%. One-tenth disagree with the statement and 12% don't know (Table 9.12.2).

Table 9.12.2 Statement: 'You should always cover food before putting it back in the fridge'

%	Agree	Disagree	Don't know
France	79	10	12
Four-country average	83	6	12

Almost all French children (96%) agree that 'You should always wash fruit and vegetables before eating them', a figure identical to the four-country average (Table 9.12.3).

Table 9.12.3 Statement: 'You should always wash fruit and vegetables before eating them'

%	Agree	Disagree	Don't know
France	96	3	2
Four-country average	96	2	3

Most French children (86%) agree that 'Food can go "off" from being kept at the wrong temperature' (Table 9.12.4). This figure is the same as the four-country average.

Table 9.12.4 Statement: 'Food can go "off" from being kept at the wrong temperature'

%	Agree	Disagree	Don't know
France	86	6	8
Four-country average	86	5	10

The majority of French children (83%) disagree with the statement 'You don't always need to wash your hands before you eat', a figure which is very close to the four-country average of 84% (Table 9.12.5). Around one-seventh (16%) agree with the statement, and only 1% don't know.

Table 9.12.5 Statement: 'You don't always need to wash your hands before you eat'

%	Agree	Disagree	Don't know
France	16	83	1
Four-country average	14	84	3

9.12.2 Understanding of 'use-by' dates

| Question: | *Which of these things do 'use-by' dates tell you?*

Choice:

- *You need to eat the food before that date.*
- *It's OK to eat it two weeks after the date mentioned.*
- *It's the date by which stores need to sell that food.*
- *Don't know.*

Three-quarters of French children believe that the 'use-by' date on foods mean that it should be eaten before the date given, a figure close to the four-country average of 77% (Table 9.12.6). Just under one-fifth (18%) interpret 'use-by' dates as the date by which shops must sell the food, which is higher than the four-country average of 13%. Overall, only 5% of French children say that they do not know what 'use-by' dates mean or give no answer.

Table 9.12.6 'Which of these things do "use-by" dates tell you?'

%	Eat before date	OK to eat 2 weeks after date	Store sell-by date	Don't know
France	75	4	18	5
Four-country average	77	4	13	8

9.12.3 Understanding of food poisoning

| Question: | How do you get food poisoning?

Choice:

- *From eating too much.*
- *From harmful bacteria in food.*
- *From the additives in food.*
- *From the pesticides that are used on crops.*
- *Don't know.*

Less than half of French children (43%) believe that food poisoning is contracted from bacteria in food, a figure substantially lower than the four-country average of 61% (Table 9.12.7). Just under one-third (30%) believe that 'eating too much' causes food poisoning. A further one-fifth believe that pesticides used on crops are a cause of food poisoning, and 13% attribute it to food additives. Over one-fifth (21%) don't know what causes food poisoning, which is substantially higher than the four-country average of 13%.

Table 9.12.7 'How do you get food poisoning?'

%	Bacteria	Pesticides	Additives	Eating too much	Don't know
France	43	20	13	30	21
Four-country average	61	28	14	17	13

9.12.4 Understanding of Salmonella

| Question: | What is Salmonella?

Choice:

- *Bacteria, which, when found in food, make you ill.*
- *A type of pink salmon.*

- *A common childhood disease.*
- *Don't know.*

Responses in this report were not broken down by age, sex and social class because it was found that the differences in responses given across these segments were not overly significant, and that general trends emerged right across these groups.

One-fifth of French children are aware that Salmonella is a form of harmful bacterium which can occur in food; this figure is much lower than the four-country average of just under one-half (49%) (Table 9.12.8). The majority (65%) don't know what Salmonella is; this figure is substantially higher than the four-country average of 41%.

Table 9.12.8 'What is Salmonella?'

%	Bacteria found in food	Pink salmon	Child disease	Don't know
France	20	1	14	65
Four-country average	49	3	8	41

9.13 Germany: current eating and drinking patterns

9.13.1 What children eat and drink

| Question: | *I want to talk to you about food and nutrition, but, first of all, can you tell me what you had to eat and drink for breakfast today/lunch today/evening meal yesterday?*

Each child was allowed to give details of one meal eaten during the 24 hours before the interview. He or she was able to select from breakfast today/lunch today/evening meal yesterday; interviewers kept a check that overall, a balance of coverage between the different meals was obtained in each country.

In Germany, the single most popular food at breakfast is bread (30%), and the most popular breakfast drink is cold milk (27%). Cereals are also popular, with 27% of children eating them.

At lunch, vegetables and salad are popular (52%) as are potatoes (28%) and meat (34%). Pasta is also fairly popular, with 16% of German children eating it for lunch. Children most frequently drink water (21%), followed by lemonade (16%) and cola (15%).

German children eat bread at dinner much more frequently (46%) than do children in the other countries. 45% eat meat and 27% eat vegetables and salad. Sausages are very popular with 37% of children eating them.

23% of children eat cheese at dinner. Cola is the most popular drink (17%), followed by tea (15%), lemonade (14%), fruit juice (11%) and water (9%).

9.13.2   Meal choices – who makes the decisions?

| Question: | *Who usually decides what you eat for:* |

- *breakfast during the week;*
- *breakfast at the weekend;*
- *lunch during the week;*
- *lunch at the weekend;*
- *dinner/supper during the week;*
- *dinner/supper at the weekend.*

Choice: Father/I do/Mother/School/Other

For the above questions, each child was asked to comment on one meal. Multiple responses account for the fact that figures do not always add up to 100%.

German children follow the pattern seen in the other countries studied and claim to exert most influence on the menu at breakfast, both during the week and at weekends (Table 9.13.1). However, a key difference from the other countries is the high level of influence German children claim to have on the dinner menu (possibly due to the fact that many children do not have a hot, cooked meal in the evening). Some 63% of German children claim to make their own choices at dinner during the week, compared with the four-country average of 28%; and 60% choose their dinner at weekends, compared with the four-country average of 28%. As in other countries (except the UK), at lunch, the chief decision-maker in Germany is the mother, who makes choices on behalf of the majority of children both during the week and at weekends.

Table 9.13.1 'Who usually decides what you eat?'

%	Breakfast		Lunch		Dinner	
	I do	Mother	I do	Mother	I do	Mother
During the week						
Germany	65	37	23	69	63	47
Four-country average	70	30	24	55	28	74
During the weekend						
Germany	63	39	17	84	60	50
Four-country average	69	31	23	76	28	73

9.13.3   Meals eaten with the family

| Question: | Which meals do you usually eat with your family?

As in the other three countries, dinner is the meal that German children eat the most frequently with their families (by 'family' we mean with at least one parent present), both during the week and at weekends (Table 9.13.2). Lunch, particularly at weekends, is also often eaten with one or more parent. During the week, over half (56%) of children eat lunch with their families, whilst the vast majority (94%) eat lunch with their families at weekends. Breakfast is eaten with the family most frequently at weekends (82%). Only 48% eat breakfast with one or more parent during the week.

Table 9.13.2 'Which meals do you eat with your family?'

%	Breakfast		Lunch		Dinner	
	W/D	W/E	W/D	W/E	W/D	W/E
Germany	48	82	56	94	88	92
Four-country average	48	65	45	86	92	94

W/D = weekday, W/E = weekend

9.14   Germany: children's views on nutrition and health

9.14.1   Perceptions of food processing and its effects on nutrition and health

| Question: | I would now like to read you a number of statements. Please tell me if you agree or disagree with these statements or if you don't know:

Statements:

- 'Ready-made meals are just as good for you as home-made meals'
- 'Tinned fruit and vegetables are just as good for you as fresh fruit and vegetables'
- 'Fresh foods are safer than frozen or tinned foods'
- 'It is safer to drink milk straight from the cow than the milk one buys in the store'
- 'The food we eat today is safer than ever before'
- 'Foods containing E numbers are bad for you'

In Germany, over half (52%) disagree with the statement 'Ready-made meals are just as good for you as home-made meals', compared with the four-country average of 61% (Table 9.14.1). Just under one-fifth (18%) agree with the statement and 30% don't know.

Table 9.14.1 Statement: 'Ready-made meals are just as good for you as home-made meals'

%	Agree	Disagree	Don't know
Germany	18	52	30
Four-country average	23	61	17

German children, in comparison to those from the other three countries, answered 'don't know' most often in response to the statement 'Tinned fruit and vegetables are just as good for you as fresh fruit and vegetables' (Table 9.14.2). One-half of German children disagree with this statement.

Table 9.14.2 Statement: 'Tinned fruit and vegetables are just as good for you as fresh fruit and vegetables'

%	Agree	Disagree	Don't know
Germany	20	50	30
Four-country average	23	60	17

Over half (55%) of German children agree that 'Fresh foods are safer than tinned or frozen foods', compared with the four-country average of 69% (Table 9.14.3). Around one-seventh (14%) disagree with this statement and just under one-third (31%) don't know.

Table 9.14.3 Statement: 'Fresh foods are safer than frozen or tinned foods'

%	Agree	Disagree	Don't know
Germany	55	14	31
Four-country average	69	14	18

Just under one-quarter (24%) of German children agree 'It is safer to drink milk straight from the cow than the milk one buys in the store', a figure close to the four-country average of (22%) (Table 9.14.4). Almost one-third (32%) disagree – the lowest percentage reported in any country studied (average 53%). A majority of 44% don't know.

Table 9.14.4 Statement: 'It is safer to drink milk straight from the cow than the milk one buys in the store'

%	Agree	Disagree	Don't know
Germany	24	32	44
Four-country average	22	53	26

Over one-fifth (21%) of German children agree that 'The food we eat today is safer than ever before', the lowest figure seen in any country studied, and substantially lower than the four-country average of 42% (Table 9.14.5). Overall, 29% of children disagree with this statement. One-half of German children claim they don't know.

Table 9.14.5 Statement: 'The food we eat today is safer than ever before'

%	Agree	Disagree	Don't know
Germany	21	29	50
Four-country average	42	27	31

Over one-fifth (22%) of children in Germany agree that 'Foods containing E numbers are bad for you', compared with the four-country average of 29% (Table 9.14.6). Some 8% disagree with the statement, whilst 69% don't know.

Table 9.14.6 Statement: 'Foods containing E numbers are bad for you'

%	Agree	Disagree	Don't know
Germany	22	8	69
Four-country average	29	13	58

9.14.2 How children rate the nutritional quality of foods and drinks

> Question: I have a list of foods here and I want you to show me which ones you think are 'Good for you', 'Okay for you' or 'Not so good for you'.

A three-point Nutritional Rating Scale developed by CRU was used to evaluate children's knowledge of nutrition. Since previous research indicates that children do not believe any food or drink to be 'bad for you', this Nutritional Rating Scale allows interviewees to rate the perceived nutritional value of individual food and drink products (from a prompted list) by indicating whether they believe a food to be:

- 'Good for you';
- 'OK for you';
- 'Not so good for you'.

Classifying foods, especially when involving children, sometimes poses difficulties. In this survey, as a result of pilot studies, potatoes are not included as vegetables, French fries are not included as potatoes, nor is chicken considered as meat; in children's minds, these foods were distinct and separate.

In Germany, the foods and drinks that the largest proportion of children believe to be 'Good for you' are fruit (81%), followed by vegetables (76%), bread (75%) and milk (69%) (Table 9.14.7). This is very similar to the four-country average ranking, which is led by fruit, vegetables, water and milk.

Table 9.14.7 Germany: Foods and drinks which children believe to be 'Good for you'

Rank		Percentage of children who rate it as 'Good for you'	Four-country average (%)
1	Fruit	81	85
2	Vegetables	76	82
3	Bread	75	60
4	Milk	69	72
5	Pasta	68	52
6	Cheese	64	56
7	Chicken	62	58
8	Breakfast cereal	54	49
9	Meat	52	50
9	Fish	52	59
9	Butter	52	25
12	Water	51	73
13	Pizza	50	27
14	Eggs	44	43
15	Soup	41	51
16	Margarine	35	19
17	French fries	32	19
18	Ketchup	30	14
19	Biscuits	29	18
20	Ice cream	28	15
21	Burgers	27	15
22	Fizzy drinks	26	13
23	Cakes	24	14
23	Chocolate	24	13
25	Crisps	23	12
26	Sweets	15	7
27	Sugar	14	13
28	Cider	2	6
28	Beer	2	4
30	Wine	1	4

The ranking of foods and drinks that the greatest percentage of German children believe to be 'Not so good for you' is led by wine (82%), followed by cider (79%), beer (75%) and sugar (40%) (Table 9.14.8). This compares with the four-country average 'Not so good for you' ranking, which is led by beer, followed by wine, sweets and cider.

Table 9.14.8 Germany: Foods and drinks which children believe to be 'Not so good for you'

Rank		Percentage of children who rate it as 'Not so good for you'	Four-country average (%)
1	Wine	82	74
2	Cider	79	66
3	Beer	75	78
4	Sugar	40	49
5	Sweets	37	69
6	Chocolate	33	56
7	Crisps	31	58
8	Fizzy drinks	30	59
9	Burgers	28	46
10	Biscuits	24	36
10	French fries	24	47
12	Ice cream	22	43
13	Water	20	6
13	Cakes	20	44
15	Margarine	18	38
16	Soup	17	7
17	Breakfast cereal	16	12
17	Eggs	16	13
19	Fish	14	8
19	Ketchup	14	47
21	Butter	13	38
22	Meat	12	8
23	Cheese	10	8
24	Vegetables	8	3
24	Pizza	8	25
26	Milk	6	3
27	Chicken	5	5
28	Pasta	4	8
29	Bread	2	4
29	Fruit	2	1

9.14.3   Attitudes towards health and nutrition

| Question: | *I would now like to read you a number of statements. Can*

you tell me if you agree or disagree with these statements, or if you don't know?

Statements:

- *'It is best to eat small amounts of different foods rather than a lot of the same food'*
- *'To stay healthy, you should eat less fat'*
- *'Milk is good for strong bones'*
- *'Foods like sweets and ice cream are OK to eat, but not all the time'*
- *'It is important to eat foods like whole grain bread and cereals'*
- *'Fast foods are OK to eat every day'*

- *'The food you eat affects your health as you grow up'*
- *'Exercise is just as important as the food you eat for staying healthy'*
- *'Chocolate is OK to eat every day'*

Just over two-thirds (67%) of German children agree that 'It is best to eat small amounts of different foods, rather than a lot of the same food', compared with the four-country average of 74% (Table 9.14.9). Some 6% of all German children disagree with the statement, and over one-quarter (27%) don't know.

Table 9.14.9 Statement: 'It is best to eat small amounts of different foods rather than a lot of the same food'

%	Agree	Disagree	Don't know
Germany	67	6	27
Four-country average	74	11	15

Over three-quarters (76%) of German children agree that 'To stay healthy, you should eat less fat', the lowest figure recorded across the four countries (average 82%) (Table 9.14.10). Some 4% of German children disagree with this statement, and 19% don't know.

Table 9.14.10 Statement: 'To stay healthy, you should eat less fat'

%	Agree	Disagree	Don't know
Germany	76	4	19
Four-country average	82	7	11

Over three-quarters (77%) of children in Germany agree that 'Milk is good for strong bones', the lowest figure reported, and less than the four-country average of 84% (Table 9.14.11). Some 4% disagree with the statement, and almost one-fifth (19%) don't know.

Table 9.14.11 Statement: 'Milk is good for strong bones'

%	Agree	Disagree	Don't know
Germany	77	4	19
Four-country average	84	4	12

Amongst German children, 82% – the same percentage as the four-country average – agree that 'Foods like sweets and ice cream are OK to eat, but not all the time' (Table 9.14.12). Under one-tenth (8%) disagree and one-tenth don't know.

Table 9.14.12 Statement: 'Foods like sweets and ice cream are OK to eat, but not all the time'

%	Agree	Disagree	Don't know
Germany	82	8	10
Four-country average	82	12	6

In line with the results obtained in the other three countries, most German children (78%) agree that 'It is important to eat foods like whole grain bread and cereals' (Table 9.14.13). Some 7% disagree with the statement, and 15% don't know.

Table 9.14.13 Statement: 'It is important to eat foods like whole grain bread and cereals'

%	Agree	Disagree	Don't know
Germany	78	7	15
Four-country average	79	10	12

Over two-thirds of German children (69%) disagree with the statement that 'Fast foods are OK to eat every day', notably less than the four-country average of 80% (Table 9.14.14). Some 6% agree with the statement, and just under one-quarter (24%) don't know.

Table 9.14.14 Statement: 'Fast foods are OK to eat every day'

%	Agree	Disagree	Don't know
Germany	6	69	24
Four-country average	8	80	13

Just over two-thirds (67%) of German children agree that 'The food you eat affects your health as you grow up', the lowest figure seen in any of the four countries surveyed (average 81%) (Table 9.14.15). Just under one-fifth (19%) disagree with the statement, and 14% don't know.

Table 9.14.15 Statement: 'The food you eat affects your health as you grow up'

%	Agree	Disagree	Don't know
Germany	67	19	14
Four-country average	81	9	10

Just over two-thirds (67%) of German children agree that 'Exercise is just as important as the food you eat for staying healthy', the lowest figure reported for the countries covered by this study, and notably below the four-country average of 79% (Table 9.14.16). Overall, 8% disagree with this statement, and the remaining 24% don't know.

Table 9.14.16 Statement: 'Exercise is just as important as the food you eat for staying healthy'

%	Agree	Disagree	Don't know
Germany	67	8	24
Four-country average	79	10	11

Amongst German children, 15% agree that 'Chocolate is OK to eat every day' – the highest figure recorded in the four countries (Table 9.14.17). Just under three-quarters (73%) disagree with the statement, and 13% don't know.

Table 9.14.17 Statement: 'Chocolate is OK to eat every day'

%	Agree	Disagree	Don't know
Germany	15	73	13
Four-country average	11	82	7

9.14.4   Understanding how nutrients relate to health

| Question: | To stay healthy, which of these does your body need?

Choice: Protein, fat, minerals, vitamins, salt, fibre, pills, calories, calcium, sugar, starch

As in the other three countries, children in Germany show a strong awareness of the importance of vitamins to health, with 94% recognizing their role (Table 9.14.18). Next in the ranking comes minerals (71%), followed by calcium (69%) and protein (66%). This compares with the four-country average ranking which is led by vitamins (94%), protein (82%), calcium (78%) and minerals (70%).

Table 9.14.18 Nutrients most widely recognized as being important for health

	Germany	Four-country average (%)
Vitamins	94	94
Minerals	71	70
Calcium	69	78
Protein	66	82
Fibre	50	61
Starch	45	37
Fat	39	32
Salt	37	34
Calories	36	43
Sugar	30	42

9.15   Germany: learning about food and nutrition

9.15.1   Children's interpretations of the word 'nutrition'

| Question: | What does the word 'nutrition' mean to you?

Most children understand that the word 'nutrition' is related to the health and dietary benefits of foods, with 27% claiming that nutrition connotes 'Good for health/Healthy eating'. 15% say that the word 'Has to do with food'. Only 6% of German children were unable to express an opinion on this word's meaning, higher than in Italy (3%), but much lower than in France (32%) and the UK (34%).

9.15.2   What children think they know about nutrition

| Question: | How much do you think you know/don't know about nutrition? Can you show me which of these statements applies to you the most?

Choice:

- *I know a lot about nutrition.*
- *I know some things about nutrition.*
- *I don't know much about nutrition.*
- *I don't know anything about nutrition.*

German children's claimed knowledge of nutrition follows the same trends as seen in the other countries (Table 9.15.1), with one-tenth saying that they know 'a lot' about nutrition (compared with a four-country average of 7%). A further 54% of German children claim to know 'some things'. Just under one-third (30%) know 'not much' about nutrition, whilst 5% know nothing or gave no answer.

Table 9.15.1 'How much do you think you know/don't know about nutrition?'

%	A lot	Some things	Not much	Nothing
Germany	10	54	30	5
Four-country average	7	56	30	7

9.15.3   Attitudes towards learning

| Question: | Do you think it is important that children learn about nutrition?

Overall, 83% of German children believe that it is important to learn about nutrition, a figure somewhat lower than the four-country average of 92%

(Table 9.15.2). Some 16% do not believe it important to learn about nutrition – the highest number in the four countries studied.

Table 9.15.2 'Do you think it is important that children learn about nutrition?'

%	Yes	No
Germany	83	16
Four-country average	92	9

9.15.4   Current sources of information on nutrition

| Question: | *Where do you learn about nutrition?* |

Choice: Family/friends/school teacher/television/school/magazines/advertising

A distinction was made between learning about nutrition at school and learning about it from a teacher because children perceive the two as being somewhat different; the child could actually learn from the teacher, or from educational materials such as books, leaflets, etc. available at school.

In common with the other countries studied, most children in Germany (78%) learn about nutrition from their families (Table 9.15.3), a figure substantially higher than the four-country average of just over two-thirds (67%). Some 41% obtain nutritional information from their school, and 29% from their school teacher. Just under one-third (30%) of German children obtain nutritional information from television programmes – the highest figure recorded across the four countries.

Table 9.15.3 'Where do you learn about nutrition?'

%	Family	School	Teacher	Television	Advertising	Magazines	Friends
Germany	78	41	29	30	20	15	13
Four-country average	67	41	34	17	12	8	4

9.15.5   Preferred future methods of learning about nutrition

| Question: | *How would you like to learn about nutrition?* |

Choice: Cooking classes at school/television programmes/magazines/special information packs for use at school

Over half (52%) of German children would, in the future, like to obtain information about nutrition from cooking classes at school – a figure comparable to the four-country average of 50% (Table 9.15.4). Just under one-quarter (23%) would prefer to learn from special information packs for use at school, a figure substantially below the four-country average of 43% and the lowest recorded in any country. Just under one-third (31%) favoured television as a source of information.

Table 9.15.4 'How would you like to learn about nutrition?'

%	Cooking classes	Information packs/leaflets	TV programmes
Germany	52	23	31
Four-country average	50	43	31

9.16 Germany: food safety and hygiene

9.16.1 Understanding of food hygiene

| Question: | *I would now like to read you a number of statements. Please tell me if you agree or disagree with these statements or if you don't know.*

Statements:

- *'It is important to keep the kitchen clean'*
- *'You should always cover food before putting it back in the fridge'*
- *'You should always wash fruit and vegetables before eating them'*
- *'Foods can go "off" from being kept at the wrong temperature'*
- *'You don't always need to wash your hands before you eat'*

The vast majority (91%) of German children agree that 'It is important to keep the kitchen clean'; the lowest figure reported, but still close to the four-country average of 95% (Table 9.16.1). Only 2% disagree with the statement, whilst 7% don't know.

Table 9.16.1 Statement: 'It is important to keep the kitchen clean'

%	Agree	Disagree	Don't know
Germany	91	2	7
Four-country average	95	2	4

Most German children (78%) agree that 'You should always cover food before putting it back in the fridge', the lowest figure seen in any country (average figure 83%) (Table 9.16.2). Some 5% disagree with the statement, whilst 17% don't know.

Table 9.16.2 Statement: 'You should always cover food before putting it back in the fridge'

%	Agree	Disagree	Don't know
Germany	78	5	17
Four-country average	83	6	12

The vast majority of German children (96%) agree that 'You should always wash fruit and vegetables before eating them', a figure identical to the four-country average (Table 9.16.3). Only 1% disagree with the statement, and 3% don't know.

Table 9.16.3 Statement: 'You should always wash fruit and vegetables before eating them'

%	Agree	Disagree	Don't know
Germany	96	1	3
Four-country average	96	2	3

Most German children (89%) agree that 'Food can go "off" from being kept at the wrong temperature', a figure close to the four-country average of 86% (Table 9.16.4). Some 2% disagree with the statement, whilst just under one-tenth (9%) don't know.

Table 9.16.4 Statement: 'Food can go "off" from being kept at the wrong temperature'

%	Agree	Disagree	Don't know
Germany	89	2	9
Four-country average	86	5	10

The majority of German children (79%) disagree with the statement 'You don't always need to wash your hands before you eat', compared with the four-country average of 84% (Table 9.16.5). Some 16% agree with the statement, and 5% don't know.

Table 9.16.5 Statement: 'You don't always need to wash your hands before you eat'

%	Agree	Disagree	Don't know
Germany	16	79	5
Four-country average	14	84	3

9.16.2   Understanding of 'use-by' dates

| Question: | Which of these things do 'use-by' dates tell you?

Choice:

- *You need to eat the food before that date.*
- *It's OK to eat it two weeks after the date mentioned.*
- *It's the date by which stores need to sell that food.*
- *Don't know.*

Over two-thirds (67%) of German children believe that the 'use-by' date means that 'You need to eat the food before that date', the lowest figure recorded in any country, compared with the four-country average of 77% (Table 9.16.6). Over one-fifth (22%) believe that it means 'The date by which stores need to sell that food', and just under one-tenth (9%) don't know.

Table 9.16.6 'Which of these things do "use-by" dates tell you?'

%	Eat before date	OK to eat 2 weeks after date	Store sell-by date	Don't know
Germany	67	8	22	9
Four-country average	77	4	13	8

9.16.3   Understanding of food poisoning

| Question: | How do you get food poisoning?

Choice:

- *From eating too much.*
- *From harmful bacteria in food.*
- *From the additives in food.*
- *From the pesticides that are used on crops.*
- *Don't know.*

In Germany, four-fifths of children – the highest figure reported in any country, and much higher than the four-country average of 61% – believe that food poisoning is caused by 'Harmful bacteria in food' (Table 9.16.7). Over one-quarter (28%) attribute it to 'The pesticides that are used on crops', 12% believe it to be caused by additives and another 12% believe it to be caused by eating too much. Less than one-tenth (9%) don't know.

Table 9.16.7 'How do you get food poisoning?'

%	Bacteria	Pesticides	Additives	Eating too much	Don't know
Germany	80	28	12	12	9
Four-country average	61	28	14	17	13

9.16.4   Understanding of Salmonella

> **Question:**   *What is Salmonella?*

Choice:

- *Bacteria, which, when found in food, make you ill.*
- *A type of pink salmon.*
- *A common childhood disease.*
- *Don't know.*

Responses in this report were not broken down by age, sex and social class because it was found that the differences in responses given across these segments were not overly significant, and that general trends emerged right across these groups.

   Some 69% of German children – the highest figure recorded in any country, and substantially higher than the four-country average of 49% – are aware that Salmonella is a bacterium which, when found in food, can cause food poisoning (Table 9.16.8). Only 1% say it is a childhood disease; a further 29% don't know.

Table 9.16.8 'What is Salmonella?'

%	Bacteria found in food	Pink salmon	Child disease	Don't know
Germany	69	0	1	29
Four-country average	49	3	8	41

9.17   Italy: current eating and drinking patterns

9.17.1   What children eat and drink

> **Question:**   *I want to talk to you about food and nutrition, but, first of all, can you tell me what you had to eat and drink for breakfast today/lunch today/evening meal yesterday?*

Each child was allowed to give details of one meal eaten during the 24 hours before the interview. He or she was able to select from breakfast today/lunch today/evening meal yesterday; interviewers kept a check that overall, a balance of coverage between the different meals was obtained in each country.

In Italy, 19% of children do not eat anything for breakfast, the highest rating of all the countries. The most popular breakfast food is bakery products (62%) with only 8% eating cereals. Most children drink milk, either hot (26%) or cold (28%). 17% drink coffee and 21% do not drink anything at breakfast.

At lunch, pastas dominate (60%), as well as vegetables and salad (49%) and meat at 40%. Only 5% eat sandwiches. 30% eat fruit, a figure which is much higher than those from the UK (5%) and Germany (2%). Most children (90%) drink water with lunch and 9% drink cola.

At dinner, meat (24%), pastas (27%) and vegetables and salad (39%) are once again popular. Also for dinner, fruit is much more popular in Italy than elsewhere, with 26% having it at this meal. 14% of children eat cheese. 81% of children choose water and 10% drink cola.

9.17.2   Meal choices – who makes the decisions?

Question: *Who usually decides what you eat for:*

- *breakfast during the week;*
- *breakfast at the weekend;*
- *lunch during the week;*
- *lunch at the weekend;*
- *dinner/supper during the week;*
- *dinner/supper at the weekend.*

Choice: Father/I do/Mother/School/Other

For the above questions, each child was asked to comment on one meal. Multiple responses account for the fact that figures do not always add up to 100%.

As in the other three countries, Italian children claim to have significant influence on the choice of menu at breakfast (Table 9.17.1), with three-fifths claiming to make their own decisions on what to eat both during the week and at weekends. This compares to a four-country average of 70% for weekdays and 69% during the weekend. At lunch and dinner, the chief decision-maker is the mother. She makes choices on behalf of four-fifths or more of Italian children both during the week and at weekends, figures which are consistently higher than the four-country averages.

Table 9.17.1 'Who usually decides what you eat?'

%	Breakfast		Lunch		Dinner	
	I do	Mother	I do	Mother	I do	Mother
During the week						
Italy	60	40	15	80	19	84
Four-country average	70	30	24	55	28	74
During the weekend						
Italy	60	39	18	81	20	81
Four-country average	69	31	23	76	28	73

9.17.3   Meals eaten with the family

Question: | *Which meals do you usually eat with your family?*

Following the trends seen in the other three countries, dinner is the meal eaten most frequently as a family (by family we mean at least one parent present) in Italy, both during the week (96%) and at weekends (95%) (Table 9.17.2). Lunch, particularly at weekends (96%), is almost always eaten with one or more parent. During the week, over three-quarters (77%) of Italian children have lunch with their families, the highest figure reported, and much higher than the four-country average of 45%. Breakfast is eaten less often with the family (by 58% during the week and 68% at weekends).

Table 9.17.2 'Which meals do you eat with your family?'

%	Breakfast		Lunch		Dinner	
	W/D	W/E	W/D	W/E	W/D	W/E
Italy	58	68	77	96	96	95
Four-country average	48	65	45	86	92	94

W/D = weekday, W/E = weekend

9.18   Italy: children's views on nutrition and health

9.18.1   Perceptions of food processing and its effects on nutrition and health

Question: | *I would now like to read you a number of statements. Please tell me if you agree or disagree with these statements or if you don't know:*

Statements:

- *'Ready-made meals are just as good for you as home-made meals'*
- *'Tinned fruit and vegetables are just as good for you as fresh fruit and vegetables'*
- *'Fresh foods are safer than frozen or tinned foods'*
- *'It is safer to drink milk straight from the cow than the milk one buys in the store'*
- *'The food we eat today is safer than ever before'*
- *'Foods containing E numbers are bad for you'*

In Italy, over four-fifths (81%) of children disagree that 'Ready-made meals are just as good for you as home-made meals' (Table 9.18.1). This is the highest figure recorded in any country, and is substantially higher than the four-country average of 61%. Some 13% of Italian children agree with the statement whilst 6% don't know.

Table 9.18.1 Statement: 'Ready-made meals are just as good for you as home-made meals'

%	Agree	Disagree	Don't know
Italy	13	81	6
Four-country average	23	61	17

Almost four-fifths (78%) of Italian children disagree with the statement that 'Tinned fruit and vegetables are just as good for you as fresh fruit and vegetables', again the highest figure reported, and notably higher than the four-country average of three-fifths (Table 9.18.2). Some 15% agree with the statement – the lowest figure seen in any country. Overall, 7% don't know whether they agree or disagree with the statement.

Table 9.18.2 Statement: 'Tinned fruit and vegetables are just as good for you as fresh fruit and vegetables'

%	Agree	Disagree	Don't know
Italy	15	78	7
Four-country average	23	60	17

Over four-fifths (84%) of Italian children believe that 'Fresh foods are safer than tinned or frozen foods', the highest figure reported in any of the four countries, and substantially above the four-country average of 69% (Table 9.18.3). One-tenth disagree with this statement and 6% don't know.

Table 9.18.3 Statement: 'Fresh foods are safer than frozen or tinned foods'

%	Agree	Disagree	Don't know
Italy	84	10	6
Four-country average	69	14	18

Almost three-fifths (59%) of Italian children disagree that 'It is safer to drink milk straight from the cow than the milk one buys in the store', compared with the four-country average of 53% (Table 9.18.4). Just over one-quarter (27%) agree with the statement and 14% don't know.

Table 9.18.4 Statement: 'It is safer to drink milk straight from the cow than the milk one buys in the store'

%	Agree	Disagree	Don't know
Italy	27	59	14
Four-country average	22	53	26

Over one-third (34%) of Italian children agree that 'The food we eat today is safer than ever before', a figure lower than the four-country average (Table 9.18.5). Just under one-half (46%) of Italian children disagree with this statement – the highest figure recorded in any country – whilst one-fifth don't know.

Table 9.18.5 Statement: 'The food we eat today is safer than ever before'

%	Agree	Disagree	Don't know
Italy	34	46	20
Four-country average	42	27	31

Some 44% of Italian children agree that 'Foods containing E numbers are bad for you', the highest figure reported in any of the countries studied, and substantially higher than the four-country average of 29% (Table 9.18.6). Just under one-tenth (9%) disagree with the statement, and almost one-half (47%) don't know.

Table 9.18.6 Statement: 'Foods containing E numbers are bad for you'

%	Agree	Disagree	Don't know
Italy	44	9	47
Four-country average	29	13	58

9.18.2   How children rate the nutritional quality of foods and drinks

| Question: | *I have a list of foods here and I want you to show me which ones you think are 'Good for you', 'Okay for you' or 'Not so good for you'.*

A three-point Nutritional Rating Scale developed by CRU was used to evaluate children's knowledge of nutrition. Since previous research indicates that children do not believe any food or drink to be 'bad for you', this Nutritional Rating Scale allows interviewees to rate the perceived nutritional value of individual food and drink products (from a prompted list) by indicating whether they believe a food to be:

- 'Good for you';
- 'OK for you';
- 'Not so good for you'

In Italy, the three-point scale was amended in order to clarify further the association between health and nutrition. Here, pilot research found that the questionnaire was most clearly understood by interviewees when the three-point scale used the following three options: 'Very good for your health'; 'Good for your health'; and 'Not so good for your health'.

Classifying foods, especially when involving children, sometimes poses difficulties. In this survey, as a result of pilot studies, potatoes are not included as vegetables, french fries are not included as potatoes, nor is chicken considered as meat; in children's minds, these foods were distinct and separate.

The ranking of foods and drinks which Italian children believe to be 'Very good for your health' is led by fruits (71%), followed by vegetables (65%), water (60%) and milk (54%) (Table 9.18.7). This is exactly identical to the top four places of the four-country average 'Good for you' ranking. The ranking of foods and drinks which Italian children believe to be 'Not so good for your health' is led by sweets (88%), fizzy drinks (87%), beer (85%) and ketchup (82%) (Table 9.18.8). This differs quite a bit from the top four places of the four-country average 'Not so good for you' ranking of beer, followed by wine, sweets and cider.

9.18.3   Attitudes towards health and nutrition

| Question: | *I would now like to read you a number of statements. Can you tell me if you agree or disagree with these statements, or if you don't know?*

Statements:

- *'It is best to eat small amounts of different foods rather than a lot of the same food'*

Table 9.18.7 Italy: Foods and drinks which children believe to be 'Very good for your health'

Rank	Percentage of children who rate it as 'Very good for your health'	Four-country average (%)
1 Fruit	71	85
2 Vegetables	65	82
3 Water	60	73
4 Milk	54	72
5 Fish	46	59
6 Meat	36	50
7 Soup	35	51
8 Bread	33	60
9 Cheese	30	56
10 Pasta	29	52
10 Chicken	29	58
12 Yoghurt	27	–
13 Breakfast cereal	23	49
14 Eggs	20	43
15 Olive oil	16	–
16 Biscuits	12	18
16 Sugar	12	13
18 Pizza	11	27
19 Burgers	8	15
20 Cakes	7	14
21 Ice cream	6	15
21 Chocolate	6	13
23 Margarine	5	19
24 French fries	4	19
25 Ketchup	3	14
25 Cider	3	6
25 Butter	3	25
25 Crisps	3	12
29 Beer	2	4
29 Wine	2	4
29 Sweets	2	7
29 Fizzy drinks	2	13

- *'To stay healthy, you should eat less fat'*
- *'Milk is good for strong bones'*
- *'Foods like sweets and ice cream are OK to eat, but not all the time'*
- *'It is important to eat foods like whole grain bread and cereals'*
- *'Fast foods are OK to eat every day'*
- *'The food you eat affects your health as you grow up'*
- *'Exercise is just as important as the food you eat for staying healthy'*
- *'Chocolate is OK to eat every day'*

Almost four-fifths (79%) of Italian children agree that 'It is best to eat small amounts of different foods, rather than a lot of the same food', compared with the four-country average of 74% (Table 9.18.9). Overall, 13% disagree with this statement, and 8% don't know.

Table 9.18.8 Italy: Foods and drinks which children believe to be 'Not so good for your health'

Rank		Percentage of children who rate it as 'Not so good for your health'	Four-country average (%)
1	Sweets	88	69
2	Fizzy drinks	87	59
3	Beer	85	78
4	Ketchup	82	47
5	Crisps	81	58
6	Wine	78	74
7	French fries	76	47
8	Chocolate	68	56
9	Butter	57	38
9	Cider	57	66
11	Margarine	54	38
11	Cakes	54	44
13	Ice cream	51	43
14	Burgers	49	46
15	Sugar	41	49
16	Pizza	32	25
17	Biscuits	28	36
18	Olive oil	19	–
19	Breakfast cereal	17	12
19	Eggs	17	13
21	Yoghurt	10	–
22	Meat	8	8
23	Cheese	7	8
23	Pasta	7	8
25	Bread	6	4
25	Fish	6	8
27	Chicken	5	5
28	Milk	3	3
28	Soup	3	7
30	Water	2	6
30	Vegetables	2	3
32	Fruit	0	1

Table 9.18.9 Statement: 'It is best to eat small amounts of different foods rather than a lot of the same food'

%	Agree	Disagree	Don't know
Italy	79	13	8
Four-country average	74	11	15

Over four-fifths (85%) of Italian children agree that 'To stay healthy, you should eat less fat', a figure slightly higher than the four-country average of 82% (Table 9.18.10). Overall, 9% of Italian children disagree with this statement, and 6% don't know.

Table 9.18.10 Statement: 'To stay healthy, you should eat less fat'

%	Agree	Disagree	Don't know
Italy	85	9	6
Four-country average	82	7	11

The majority of Italian children (88%) agree that 'Milk is good for strong bones', compared with the four-country average of 84% (Table 9.18.11). Some 4% disagree and 8% don't know.

Table 9.18.11 Statement: 'Milk is good for strong bones'

%	Agree	Disagree	Don't know
Italy	88	4	8
Four-country average	84	4	12

Over four-fifths (81%) of Italian children agree that 'Foods like sweets and ice cream are OK to eat, but not all the time', compared with the four-country average of 82% (Table 9.18.12). Some 17% of Italian children disagree with the statement and 2% don't know.

Table 9.18.12 Statement: 'Foods like sweets and ice cream are OK to eat, but not all the time'

%	Agree	Disagree	Don't know
Italy	81	17	2
Four-country average	82	12	6

Some 69% of Italian children agree that 'It is important to eat foods like whole grain bread and cereals', the lowest figure recorded across the four countries, and less than the four-country average of 79% (Table 9.18.13). Overall, 18% of Italian children disagree with the statement and 13% don't know.

Table 9.18.13 Statement: 'It is important to eat foods like whole grain bread and cereals'

%	Agree	Disagree	Don't know
Italy	69	18	13
Four-country average	79	10	12

Only 5% of Italian children agree that 'Fast foods are okay to eat every day', compared with the four-country average of 8%. This is the lowest figure reported for any country studied (Table 9.18.14). Overall, 85% disagree with the statement while one-tenth of Italian children don't know.

Table 9.18.14 Statement: 'Fast foods are OK to eat every day'

%	Agree	Disagree	Don't know
Italy	5	85	10
Four-country average	8	80	13

In Italy, the vast majority (94%) of children agree that 'The food you eat affects your health as you grow up', compared with the four-country average of 81% (Table 9.18.15). This figure was the highest from any country surveyed. Some 4% of children disagree with the statement and 2% don't know.

Table 9.18.15 Statement: 'The food you eat affects your health as you grow up'

%	Agree	Disagree	Don't know
Italy	94	4	2
Four-country average	81	9	10

Four-fifths of Italian children agree that 'Exercise is just as important as the food you eat for staying healthy', a figure which is close to the four-country average of 79% (Table 9.18.16). Some 15% disagree with this statement, and 5% don't know.

Table 9.18.16 Statement: 'Exercise is just as important as the food you eat for staying healthy'

%	Agree	Disagree	Don't know
Italy	80	15	5
Four-country average	79	10	11

The majority (88%) of children in Italy disagree with the statement 'Chocolate is OK to eat every day', the highest figure reported in any country, and somewhat higher than the four-country average of 82% (Table 9.18.17). Some 8% agree with the statement, and 4% don't know.

Table 9.18.17 Statement: 'Chocolate is OK to eat every day'

%	Agree	Disagree	Don't know
Italy	8	88	4
Four-country average	11	82	7

9.18.4   Understanding how nutrients relate to health

| Question: | To stay healthy, which of these does your body need?

Choice: Protein, fat, minerals, vitamins, salt, fibre, pills, calories, calcium, sugar, starch

In Italy, as in all the other countries studied, vitamins are most widely acknowledged by children as being important for health (93% of children agree that these are needed for staying healthy), along with protein (93%), followed by calcium (82%) and sugar (53%). Sugar, however, does not appear in the top four rankings from any other country (Table 9.18.18). Across the four countries, vitamins lead the ranking, followed by protein, calcium and minerals.

Table 9.18.18 Nutrients most widely recognized as being important for health

	Italy	Four-country average (%)
Vitamins	93	94
Protein	93	82
Calcium	82	78
Sugar	53	42
Minerals	51	70
Fibre	51	61
Calories	44	43
Starch	23	38
Fat	23	31
Salt	18	34

9.19   Italy: learning about food and nutrition

9.19.1   Children's interpretations of the word 'nutrition'

| Question: | What does the word 'nutrition' mean to you?

76% of Italian children understand that the word 'nutrition' is related to the health and dietary benefits of foods; 12% agree that nutrition connotes 'Good for health/Healthy eating'. Only 2% say that the word 'Has to do with food'. Only 3% of Italian children were unable to offer an opinion as to the word's meaning. This compares to figures of 6% in Germany, 32% in France and 34% in the UK.

9.19.2 What children think they know about nutrition

| Question: | How much do you think you know/don't know about nutrition? Can you show me which of these statements applies to you the most?

Choice:

- *I know a lot about nutrition.*
- *I know some things about nutrition.*
- *I don't know much about nutrition.*
- *I don't know anything about nutrition.*

Only 5% of children in Italy claim to 'know a lot' about nutrition; this figure is similar to the four-country average of 7% (Table 9.19.1). Two-thirds of Italian children claim to 'know some things', whereas, in the four countries surveyed, an average of 56% said the same thing. Over one-quarter (27%) of Italian children 'don't know much', which is nearly equivalent to the survey average of 30%. Some 3% of children in Italy 'don't know anything' which is lower than the overall 7% average.

Table 9.19.1 'How much do you think you know/don't know about nutrition?'

%	A lot	Some things	Not much	Nothing
Italy	5	66	27	3
Four-country average	7	56	30	7

9.19.3 Attitudes towards learning

| Question: | Do you think it is important that children learn about nutrition?

The vast majority (97%) of children in Italy believe it important that they learn about nutrition (Table 9.19.2), which constitutes the highest figure recorded in any of the four countries (average 92%). Only 3% do not believe it important to learn about nutrition.

Table 9.19.2 'Do you think it is important that children learn about nutrition?'

%	Yes	No
Italy	97	3
Four-country average	92	9

9.19.4   Current sources of information on nutrition

| Question: | *Where do you learn about nutrition?*

Choice: Family/friends/school teacher/television/school/magazines/advertising

A distinction was made between learning about nutrition at school and learning about it from a teacher because children perceive the two as being somewhat different; the child could actually learn from the teacher, or from educational materials such as books, leaflets, etc. available at school.

Over four-fifths (81%) of Italian children say that they learn about nutrition from their families – the highest figure reported from any country, and substantially higher than the four-country average of 67% (Table 9.19.3). The child's school (38%) and school teachers (37%) also play a part. Television programmes are a source of information for 11% of children.

Table 9.19.3 'Where do you learn about nutrition?'

%	Family	School	Teacher	Television	Advertising	Magazines	Friends
Italy	81	38	37	11	7	8	3
Four-country average	67	41	34	17	12	8	4

9.19.5   Preferred future methods of learning about nutrition

| Question: | *How would you like to learn about nutrition?*

Choice: Cooking classes at school/television programmes/magazines/special information packs for use at school

In contrast with the findings from other countries, the greatest percentage (68%) of Italian children would like to learn more about nutrition from information packs for use at school, with just under one-half (49%) preferring cooking classes at school and 41% opting for television programmes (Table 9.19.4). The Italian results compare with the following four-country averages: 50% would like to learn from cooking classes, 43% would choose information packs for use at school and almost one-third would prefer TV programmes.

Table 9.19.4 'How would you like to learn about nutrition?'

%	Cooking classes	Information packs/leaflets	TV programmes
Italy	49	68	41
Four-country average	50	43	31

Multiple responses allowed.

9.20   Italy: food safety and hygiene

9.20.1   Understanding of food hygiene

| Question: | *I would now like to read you a number of statements. Please tell me if you agree or disagree with these statements or if you don't know.*

Statements:

- *'It is important to keep the kitchen clean'*
- *'You should always cover food before putting it back in the fridge'*
- *'You should always wash fruit and vegetables before eating them'*
- *'Foods can go "off" from being kept at the wrong temperature'*
- *'You don't always need to wash your hands before you eat'*

The vast majority (99%) of Italian children agree that 'It is important to keep the kitchen clean', the highest figure recorded in any country (average 95%), with only 0.5% disagreeing and 0.5% not knowing (Table 9.20.1).

Table 9.20.1 Statement: 'It is important to keep the kitchen clean'

%	Agree	Disagree	Don't know
Italy	99	0.5	0.5
Four-country average	95	2	4

The majority (86%) of children in Italy agree that 'You should always cover food before putting it back in the fridge', a figure slightly higher than the four-country average of 83% (Table 9.20.2).

Table 9.20.2 Statement: 'You should always cover food before putting it back in the fridge'

%	Agree	Disagree	Don't know
Italy	86	4	10
Four-country average	83	6	12

Almost all Italian children (99%) agree that 'You should always wash fruit and vegetables before eating them', the highest figure reported from any of the four countries surveyed (average 96%). Some 1% disagree with the statement and 0% 'don't know' (Table 9.20.3).

Table 9.20.3 Statement: 'You should always wash fruit and vegetables before eating them'

%	Agree	Disagree	Don't know
Italy	99	1	0
Four-country average	96	2	3

Most Italian children (83%) agree that 'Food can go "off" from being kept at the wrong temperature', a figure close to the four-country average of 86% (Table 9.20.4). Some 7% disagree with the statement and one-tenth don't know.

Table 9.20.4 Statement: 'Food can go "off" from being kept at the wrong temperature'

%	Agree	Disagree	Don't know
Italy	83	7	10
Four-country average	86	5	10

The majority of Italian children (92%) disagree with the statement 'You don't always need to wash your hands before you eat', the highest figure recorded in the four countries studied (average 84%), whilst 8% agree with it (Table 9.20.5).

Table 9.20.5 Statement: 'You don't always need to wash your hands before you eat'

%	Agree	Disagree	Don't know
Italy	8	92	0
Four-country average	14	84	3

9.20.2   Understanding of 'use-by' dates

Question: | *Which of these things do 'use-by' dates tell you?*

Choice:

- *You need to eat the food before that date.*
- *It's OK to eat it two weeks after the date mentioned.*
- *It's the date by which stores need to sell that food.*
- *Don't know.*

The vast majority (96%) of Italian children are aware that 'use-by' dates on food mean that 'You need to eat the food before that date', the greatest proportion seen in any country and substantially higher than the four-country average of 77% (Table 9.20.6).

Table 9.20.6 'Which of these things do "use-by" dates tell you?'

%	Eat before date	OK to eat 2 weeks after date	Store sell-by date	Don't know
Italy	96	0	2	2
Four-country average	77	4	13	8

9.20.3 Understanding of food poisoning

| Question: | How do you get food poisoning?

Choice:

- *From eating too much.*
- *From harmful bacteria in food.*
- *From the additives in food.*
- *From the pesticides that are used on crops.*
- *Don't know.*

In Italy, 56% of children believe that food poisoning is contracted from harmful bacteria in food, with one-half believing pesticides to be responsible, 28% attributing it to additives and 24% to eating too much (Table 9.20.7). The greatest difference from the trends in other countries is seen with the high level of belief that pesticides cause food poisoning – this figure of 50% is far higher than that seen in any other country, and notably above the four-country average of 28%. The percentage of children who believe that food poisoning is contracted from food additives (28%) is double the four-country average.

Table 9.20.7 'How do you get food poisoning?'

%	Bacteria	Pesticides	Additives	Eating too much	Don't know
Italy	56	50	28	24	4
Four-country average	61	28	14	17	13

9.20.4 Understanding of Salmonella

| Question: | What is Salmonella?

Choice:

- *Bacteria, which, when found in food, make you ill.*
- *A type of pink salmon.*
- *A common childhood disease.*
- *Don't know.*

Responses in this report were not broken down by age, sex and social class because it was found that the differences in responses given across these segments were not overly significant, and that general trends emerged right across these groups.

56% of Italian children are aware that Salmonella is a bacterium which, when found in food, can make you ill (Table 9.20.8). This figure compares with the four-country average of 49%.

Table 9.20.8 'What is Salmonella?'

%	Bacteria found in food	Pink salmon	Child disease	Don't know
Italy	56	6	14	24
Four-country average	49	3	8	41

9.21   UK: current eating and drinking patterns

9.21.1   What children eat and drink

| Question: | *I want to talk to you about food and nutrition, but, first of all, can you tell me what you had to eat and drink for breakfast today/lunch today/evening meal yesterday?*

Each child was allowed to give details of one meal eaten during the 24 hours before the interview. He or she was able to select from breakfast today/lunch today/evening meal yesterday; interviewers kept a check that overall, a balance of coverage between the different meals was obtained in each country.

In the UK, most children (57%) eat cereals for breakfast, whilst 26% have toast. Tea is the most popular drink (26%), followed by cold milk at 14%. 9% of children in the UK do not eat or drink anything for breakfast.

At lunch time, potatoes are the single most popular food (41%), with 50% also having vegetables and salad. Meat is also popular (31%), and, exclusive to the UK, sandwiches are very popular at 24%. The most popular drink at lunch time is cola (22%), followed by water (15%) and orange squash (12%).

At the evening meal, potatoes are again the single most popular food (56%), with 68% of those surveyed eating vegetables and salad. Poultry (14%) and fish (8%) were not that popular. Both water and tea are preferred by 14% of UK children, cola by 12%.

French fries in the UK proved to be extremely popular at both lunch (30%) and dinner (34%). These figures are much higher than those in other countries (the next highest scores for fries are found in France where 15% have them for lunch and 9% for dinner).

9.21.2 Meal choices – who makes the decisions?

Question:	Who usually decides what you eat for:

- *breakfast during the week;*
- *breakfast at the weekend;*
- *lunch during the week;*
- *lunch at the weekend;*
- *dinner/supper during the week;*
- *dinner/supper at the weekend.*

Choices: Father/I do/Mother/School/Other

For the above questions, each child was asked to comment on one meal. Multiple responses account for the fact that figures do not always add up to 100%.

As in the other three countries, UK children claim to have a significant influence on the choice of menu at breakfast (Table 9.21.1), with over three-quarters (the highest figures recorded) choosing the menu both during the week (78%) and at weekends (77%). Across the four countries studied, 70% claimed to make their own decisions on what to eat at breakfast on weekdays, whilst at weekends, an average of 69% of children choose their breakfast menu. At lunch in the UK, the mother has a much smaller influence on menu choice than in the other countries surveyed, with just under one-half of UK children selecting their menus both during the week and at weekends, compared with four-country averages of close to one-quarter. At dinner in the UK, as in the other countries, the chief decision-maker is the mother, who makes choices on behalf of almost four-fifths of children both during the week and at weekends.

Table 9.21.1 'Who usually decides what you eat?'

%	Breakfast		Lunch		Dinner	
	I do	Mother	I do	Mother	I do	Mother
During the week						
UK	78	20	49	29	19	79
Four-country average	70	30	24	55	28	74
During the weekend						
UK	77	21	45	55	18	77
Four-country average	69	31	23	76	28	73

9.21.3   Meals eaten with the family

| Question: | *Which meals do you usually eat with your family?* |

In the UK, children eat fewer meals with their families (by family, we mean at least one parent present) than in any of the other countries surveyed, both during the week and at weekends (Table 9.21.2). As in the other countries, dinner is the meal eaten most frequently with the family, both during the week and at weekends. Lunch, particularly at weekends, is fairly frequently eaten with one or more parent. During the week, only 4% of UK children eat lunch with their families, compared with the four-country average of 45%. Breakfast is eaten less often with the family in the UK than elsewhere both during the week (36% compared with the four-country average of 48%) and at weekends (47% compared with the four-country average of 65%).

Table 9.21.2 'Which meals do you eat with your family?'

%	Breakfast		Lunch		Dinner	
	W/D	W/E	W/D	W/E	W/D	W/E
UK	36	47	4	58	87	91
Four-country average	48	65	45	86	92	94

W/D = weekday, W/E = weekend

9.22   UK: children's views on nutrition and health

9.22.1   Perceptions of food processing and its effects on nutrition and health

| Question: | *I would now like to read you a number of statements. Please* |

tell me if you agree or disagree with these statements or if you don't know:

Statements:

- *'Ready-made meals are just as good for you as home-made meals'*
- *'Tinned fruit and vegetables are just as good for you as fresh fruit and vegetables'*
- *'Fresh foods are safer than frozen or tinned foods'*
- *'It is safer to drink milk straight from the cow than the milk one buys in the store'*
- *'The food we eat today is safer than ever before'*
- *'Foods containing E numbers are bad for you'*

Of the four countries, the highest level of agreement that 'Ready-made meals are just as good for you as home-made meals' was seen amongst UK children, at over 34% (Table 9.22.1). This figure is notably higher than the four-country average of 23%. Some 43% of UK children disagree with the statement and just under one-quarter (24%) don't know.

Table 9.22.1 Statement: 'Ready-made meals are just as good for you as home-made meals'

%	Agree	Disagree	Don't know
UK	34	43	24
Four-country average	23	61	17

One-third of UK children agree that 'Tinned fruit and vegetables are just as good for you as fresh fruit and vegetables' (Table 9.22.2). This is the highest figure reported from the four countries, and substantially higher than the average of 23%. Some 45% of UK children disagree with the statement whilst 23% don't know.

Table 9.22.2 Statement: 'Tinned fruit and vegetables are just as good for you as fresh fruit and vegetables'

%	Agree	Disagree	Don't know
UK	33	45	23
Four-country average	23	60	17

Less than two-thirds (63%) of children in the UK agree that 'Fresh foods are safer than tinned or frozen foods', a figure comparable to the four-country average of 69% (Table 9.22.3). Some 14% disagree with the statement and just under one-quarter (24%) don't know.

Table 9.22.3 Statement: 'Fresh foods are safer than frozen or tinned foods'

%	Agree	Disagree	Don't know
UK	63	14	24
Four-country average	69	14	18

The strongest disagreement with the statement that 'It is safer to drink milk straight from the cow than the milk one buys in the store' was seen in the UK, at 69% compared with the four-country average of 53% (Table 9.22.4). UK children also show the lowest level of agreement with the statement, at only 8% compared with the average figure of 22%. Just under one-quarter (23%) don't know.

Table 9.22.4 Statement: 'It is safer to drink milk straight from the cow than the milk one buys in the store'

%	Agree	Disagree	Don't know
UK	8	69	23
Four-country average	22	53	26

Over half (51%) of UK children agree that 'The food we eat today is safer than ever before', compared with the four-country average of 42% (Table 9.22.5). Some 14% disagree with the statement, substantially lower than the four-country average of 27%. Slightly over one-third (35%) don't know.

Table 9.22.5 Statement: 'The food we eat today is safer than ever before'

%	Agree	Disagree	Don't know
UK	51	14	35
Four-country average	42	27	31

Less than one-fifth (18%) of UK children disagree that 'Foods containing E numbers are bad for you'. The UK figure is higher than the four-country average of 13% (Table 9.22.6). A number identical to the four-country average (29%) agree with the statement. Just over half don't know whether they agree or disagree with the statement.

Table 9.22.6 Statement: 'Foods containing E numbers are bad for you'

%	Agree	Disagree	Don't know
UK	29	18	52
Four-country average	29	13	58

9.22.2   How children rate the nutritional quality of foods and drinks

| Question: | *I have a list of foods here and I want you to show me which ones you think are 'Good for you', 'Okay for you' or 'Not so good for you'.*

A three-point Nutritional Rating Scale developed by CRU was used to evaluate children's knowledge of nutrition. Since previous research indicates that children do not believe any food or drink to be 'bad for you', this Nutritional Rating Scale allows interviewees to rate the perceived nutritional value of individual food and drink products (from a prompted list) by indicating whether they believe a food to be:

- 'Good for you'
- 'OK for you'
- 'Not so good for you'

Classifying foods, especially when involving children, sometimes poses difficulties. In this survey, as a result of pilot studies, potatoes are not included as vegetables, French fries are not included as potatoes, nor is chicken considered as meat; in children's minds, these foods were distinct and separate.

In the view of UK children, the ranking of foods and drinks which are 'Good for you' is led by vegetables and fruit (both 94%), water (84%) and milk (75%). This is almost identical to the four-country average ranking, which is led by fruit, followed by vegetables, water and milk (Table 9.22.7).

Table 9.22.7 UK: Foods and drinks which children believe to be 'Good for you'

Rank		Percentage of children who rate it as 'Good for you'	Four-country average (%)
1	Vegetables	94	82
1	Fruit	94	85
3	Water	84	73
4	Milk	75	72
5	Fish	61	59
6	Chicken	59	58
7	Bread	58	60
8	Eggs	47	43
8	Pasta	47	52
10	Cheese	44	56
11	Breakfast cereal	43	49
12	Meat	39	50
12	Soup	39	51
14	Margarine	16	19
15	Butter	14	25
15	Pizza	14	27
17	Burgers	11	15
18	French fries	8	19
18	Ketchup	8	14
20	Crisps	7	12
20	Ice cream	7	15
20	Fizzy drinks	7	13
23	Biscuits	6	18
23	Cakes	6	14
23	Beer	6	4
26	Wine	5	4
26	Cider	5	6
28	Sugar	4	13
28	Chocolate	4	13
30	Sweets	3	7

The UK 'Not so good for you' ranking is led by sweets (84%), followed by chocolate (80%), sugar (75%) and beer (72%) (Table 9.22.8). This differs substantially from the four-country average ranking of beer, followed by wine, sweets and cider.

9.22.3 Attitudes towards health and nutrition

| Question: | *I would now like to read you a number of statements. Can you tell me if you agree or disagree with these statements, or if you don't know?*

Statements:

- *'It is best to eat small amounts of different foods rather than a lot of the same food'*

Table 9.22.8 UK: Foods and drinks which children believe to be 'Not so good for you'

Rank	Percentage of children who rate it as 'Not so good for you'	Four-country average (%)
1 Sweets	84	69
2 Chocolate	80	56
3 Sugar	75	49
4 Beer	72	78
5 Cider	69	66
6 Cakes	68	44
7 Fizzy drinks	66	59
8 Crisps	65	58
9 Wine	63	74
10 French fries	61	47
11 Biscuits	60	36
12 Ice cream	54	43
13 Butter	50	38
13 Burgers	50	46
15 Ketchup	42	47
16 Margarine	37	38
17 Pizza	36	25
18 Meat	11	8
18 Cheese	11	8
20 Eggs	10	13
21 Pasta	9	8
22 Breakfast cereal	8	12
23 Fish	7	8
24 Chicken	6	5
25 Bread	5	4
26 Soup	4	7
27 Milk	2	3
28 Water	1	6
29 Vegetables	0	3
29 Fruit	0	1

- *'To stay healthy, you should eat less fat'*
- *'Milk is good for strong bones'*
- *'Foods like sweets and ice cream are OK to eat, but not all the time'*
- *'It is important to eat foods like whole grain bread and cereals'*
- *'Fast foods are OK to eat every day'*
- *'The food you eat affects your health as you grow up'*
- *'Exercise is just as important as the food you eat for staying healthy'*
- *'Chocolate is OK to eat every day'*

Over two-thirds (67%) of UK children agree that 'It is best to eat small amounts of different foods, rather than a lot of the same food', compared with the four-country average of 74% (Table 9.22.9). Some 14% disagree with this statement, and 18% don't know.

Table 9.22.9 Statement: 'It is best to eat small amounts of different foods rather than a lot of the same food'

%	Agree	Disagree	Don't know
UK	67	14	18
Four-country average	74	11	15

Over four-fifths (83%) of children in the UK agree that 'To stay healthy, you should eat less fat', a figure close to the four-country average of 82% (Table 9.22.10). Some 8% of UK children disagree with this statement, and 10% don't know.

Table 9.22.10 Statement: 'To stay healthy, you should eat less fat'

%	Agree	Disagree	Don't know
UK	83	8	10
Four-country average	82	7	11

Over four-fifths (86%) of children in the UK agree that 'Milk is good for strong bones', a percentage close to the four-country average of 84% (Table 9.22.11). Overall, some 4% disagree with the statement and one-tenth don't know.

Table 9.22.11 Statement: 'Milk is good for strong bones'

%	Agree	Disagree	Don't know
UK	86	4	10
Four-country average	84	4	12

Four-fifths of UK children agree that 'Foods like sweets and ice cream are OK to eat, but not all the time' (Table 9.22.12). This is close to the four-country average of 82%. Some 11% disagree and 8% don't know.

Table 9.22.12 Statement: 'Foods like sweets and ice cream are OK to eat, but not all the time'

%	Agree	Disagree	Don't know
UK	80	11	8
Four-country average	82	12	6

Some 85% of UK children agree that 'It is important to eat foods like whole grain bread and cereals', the highest figure recorded, compared with the four-country average of 79% (Table 9.22.13). Only 4% disagree with the statement and 11% don't know.

Table 9.22.13 Statement: 'It is important to eat foods like whole grain bread and cereals'

%	Agree	Disagree	Don't know
UK	85	4	11
Four-country average	79	10	12

Some 14% of children in the UK agree that 'Fast foods are OK to eat every day', the highest figure recorded, and somewhat higher than the four-country average of 8% (Table 9.22.14). Almost three-quarters (73%) disagree with the statement. Some 13% don't know.

Table 9.22.14 Statement: 'Fast foods are OK to eat every day'

%	Agree	Disagree	Don't know
UK	14	73	13
Four-country average	8	80	13

Four-fifths of UK children agree that 'The food you eat affects your health as you grow up', compared with the four-country average of 81% (Table 9.22.15). 7% of UK children disagree and 13% don't know.

Table 9.22.15 Statement: 'The food you eat affects your health as you grow up'

%	Agree	Disagree	Don't know
UK	80	7	13
Four-country average	81	9	10

Over four-fifths (85%) of children in the UK agree that 'Exercise is just as important as the food you eat for staying healthy', a figure slightly higher than the four-country average of 79% (Table 9.22.16). Some 5% of UK children disagree with this statement, and one-tenth don't know.

Table 9.22.16 Statement: 'Exercise is just as important as the food you eat for staying healthy'

%	Agree	Disagree	Don't know
UK	85	5	10
Four-country average	79	10	11

Over four-fifths (84%) of UK children disagree with the statement 'Chocolate is OK to eat every day', which is close to the four-country average of 82% (Table 9.22.17). Some 7% agree with the statement and 9% don't know.

Table 9.22.17 Statement: 'Chocolate is OK to eat every day'

%	Agree	Disagree	Don't know
UK	7	84	9
Four-country average	11	82	7

9.22.4   Understanding how nutrients relate to health

| Question: | To stay healthy, which of these does your body need?

Choice: Protein, fat, minerals, vitamins, salt, fibre, pills, calories, calcium, sugar, starch

As in the other three countries, vitamins lead the ranking of nutrients considered important for health (91%). These are followed by protein (78%), minerals (75%) and fibre (73%). The UK was the only country in which fibre ranked in the top four (Table 9.22.18).

Table 9.22.18 Nutrients most widely recognized as being important for health

	UK	Four-country average (%)
Vitamins	91	94
Protein	78	82
Minerals	75	70
Fibre	73	61
Calcium	71	78
Calories	39	43
Starch	35	37
Sugar	24	42
Fat	23	32
Salt	22	34

9.23   UK: learning about food and nutrition

9.23.1   Children's interpretations of the word 'nutrition'

| Question: | What does the word 'nutrition' mean to you?

Most children understand that the word 'nutrition' is related to the health and dietary benefits of foods, with 35% agreeing that nutrition connotes 'Good for health/Healthy eating'. 11% say that the word 'Has to do with food'; over one-third of children in the UK (34%) are unable to express any opinion on the word's meaning, which compares to figures of 3% in Italy, 6% in Germany and 32% in France.

9.23.2   What children think they know about nutrition

| Question: | How much do you think you know/don't know about nutrition? Can you show me which of these statements applies to you the most?

Choice:

- *I know a lot about nutrition*
- *I know some things about nutrition*
- *I don't know much about nutrition*
- *I don't know anything about nutrition*

Only 5% of UK children claim to 'know a lot' about nutrition, compared with the four-country average of 7% (Table 9.23.1). Some 44% of UK children 'know some things', well below the four-country average of 56%. Over one-third (37%) of children in the UK claim that they 'don't know much' about nutrition, which is higher than the four-country average of 30%. 14% of UK children 'don't know anything' about the subject, which is double the four-country average of 7%.

Table 9.23.1 'How much do you think you know/don't know about nutrition?'

%	A lot	Some things	Not much	Nothing
UK	5	44	37	14
Four-country average	7	56	30	7

9.23.3   Attitudes towards learning

| Question: | Do you think it is important that children learn about nutrition?

The vast majority (92%) of UK children believe it important that they learn about nutrition, a figure identical to the four-country average (Table 9.23.2). Some 8% do not believe it to be important.

Table 9.23.2 'Do you think it is important that children learn about nutrition?'

%	Yes	No
UK	92	8
Four-country average	92	9

9.23.4 Current sources of information on nutrition

| Question: | *Where do you learn about nutrition?*

Choice: Family/friends/school teacher/television/school/magazines/advertising

A distinction was made between learning about nutrition at school and learning about it from a teacher because children perceive the two as being somewhat different; the child could actually learn from the teacher, or from educational materials such as books, leaflets, etc. available at school.

In sharp contrast to the other three countries surveyed, where the family is the most important source of information on nutrition, in the UK the main sources are the child's school (49%) and teacher (36%), followed by family (35%) and television (16%) (Table 9.23.3).

Table 9.23.3 'Where do you learn about nutrition?'

%	Family	School	Teacher	Television	Advertising	Magazines	Friends
UK	35	49	36	16	9	7	2
Four-country average	67	41	34	17	12	8	4

9.23.5 Preferred future methods of learning about nutrition

| Question: | *How would you like to learn about nutrition?*

Choice: Cooking classes at school/television programmes/magazines/special information packs for use at school

In the UK, over one-half (56%) of the children surveyed would, in the future, like to obtain information on nutrition from cooking classes at school (Table 9.23.4). This compares with a four-country average of 50%.

Information packs for use at school are favoured by 41% of UK children and TV programmes by one-quarter.

Table 9.23.4 'How would you like to learn about nutrition?'

%	Cooking classes	Information packs/leaflets	TV programmes
UK	56	41	25
Four-country average	50	43	31

9.24 UK: food safety and hygiene

9.24.1 Understanding of food hygiene

| Question: | *I would now like to read you a number of statements. Please tell me if you agree or disagree with these statements or if you don't know.*

Statements:

- *'It is important to keep the kitchen clean'*
- *'You should always cover food before putting it back in the fridge'*
- *'You should always wash fruit and vegetables before eating them'*
- *'Foods can go "off" from being kept at the wrong temperature'*
- *'You don't always need to wash your hands before you eat'*

In line with responses from the other countries, the vast majority (95%) of UK children agree that 'It it important to keep the kitchen clean' (Table 9.24.1). This figure is identical to the four-country average.

Table 9.24.1 Statement: 'It is important to keep the kitchen clean'

%	Agree	Disagree	Don't know
UK	95	1	4
Four-country average	95	2	4

The majority (88%) of UK children agree that 'You should always cover food before putting it back in the fridge', compared with a four-country average of 83% (Table 9.24.2).

Table 9.24.2 Statement: 'You should always cover food before putting it back in the fridge'

%	Agree	Disagree	Don't know
UK	88	3	9
Four-country average	83	6	12

Most UK children (91%) agree that 'You should always wash fruit and vegetables before eating them', a percentage slightly below the four-country average of 96% (Table 9.24.3).

Table 9.24.3 Statement: 'You should always wash fruit and vegetables before eating them'

%	Agree	Disagree	Don't know
UK	91	1	7
Four-country average	96	2	3

Some 85% of UK children agree that 'Food can go "off" from being kept at the wrong temperature', a figure close to the four-country average of 86% (Table 9.24.4). Some 12% were unable to answer.

Table 9.24.4 Statement: 'Food can go "off" from being kept at the wrong temperature'

%	Agree	Disagree	Don't know
UK	85	3	12
Four-country average	86	5	10

Four-fifths of UK children disagree with the statement 'You don't always need to wash your hands before you eat', compared with a four-country average of 84% (Table 9.24.5). Some 14% agree with the statement and 6% don't know.

Table 9.24.5 Statement: 'You don't always need to wash your hands before you eat'

%	Agree	Disagree	Don't know
UK	14	80	6
Four-country average	14	84	3

9.24.2 Understanding of 'use-by' dates

| Question: | *Which of these things do 'use-by' dates tell you?*

Choice:

- *You need to eat the food before that date*
- *It's OK to eat it two weeks after the date mentioned*
- *It's the date by which stores need to sell that food*
- *Don't know*

Some 71% of UK children are aware that 'use-by' dates on food mean that 'You need to eat the food before that date' (Table 9.24.6), compared with the four-country average of over three-quarters (77%). Some 11% believe that the 'use-by' date indicates the date by which the shop should sell the food, and 14% don't know.

Table 9.24.6 'Which of these things do "use-by" dates tell you?'

%	Eat before date	OK to eat 2 weeks after date	Store sell-by date	Don't know
UK	71	3	11	14
Four-country average	77	4	13	8

9.24.3   Understanding of food poisoning

Question: *How do you get food poisoning?*

Choice:

- *From eating too much*
- *From harmful bacteria in food*
- *From the additives in food*
- *From the pesticides that are used on crops*
- *Don't know*

Some 65% of UK children are aware that food poisoning can be contracted from 'Harmful bacteria in food', compared with a four-country average of 61% (Table 9.24.7). Some 15% attribute it to 'The pesticides that are used on crops', and 17% don't know.

Table 9.24.7 'How do you get food poisoning?'

%	Bacteria	Pesticides	Additives	Eating too much	Don't know
UK	65	15	2	2	17
Four-country average	61	28	14	17	13

9.24.4   Understanding of Salmonella

Question: *What is Salmonella?*

Choice:

- *Bacteria, which, when found in food, make you ill.*
- *A type of pink salmon.*

- *A common childhood disease.*
- *Don't know.*

Responses in this report were not broken down by age, sex and social class because it was found that the differences in responses given across these segments were not overly significant, and that general trends emerged right across these groups.

One-half of UK children agree that Salmonella is a form of harmful bacterium which, when found in food can make you ill (Table 9.24.8), this percentage is close to the four-country average of 49%. Some 44% don't know, a figure slightly higher than the four-country average of 41%.

Table 9.24.8 'What is Salmonella?'

%	Bacteria found in food	Pink salmon	Child disease	Don't know
UK	50	3	3	44
Four-country average	49	3	8	41

9.25 Technical appendix

This report is based on research done on 8–15-year-old children in France, Germany, Italy and the UK by the Children's Research Unit, in conjunction with research agencies from each country. Field work took place at the following times: November 1994 in Germany and the UK; December 1994 and January 1995 in France; and May 1995 in Italy.

Face-to-face interviews were carried out with at least 400 children at 25 locations in each country; demographic statistics were used to select regions so that each country's population as a whole was represented. For results based on samples of this size, there is 95% confidence that any error due to sampling and other random effects could be plus or minus 5%.

Research on children is highly method-sensitive: inappropriate methodologies can give inaccurate results. Particular efforts were made to establish a rapport with each child, and to communicate clearly that there are no 'right' or 'wrong' responses to questions. Face-to-face interviews were used with the aim of increasing the child's feeling of accountability.

1. France
   In France, interviews were carried out with 400 children at 25 sampling points which included two or more towns from each of nine regions: Paris and Paris Region, Paris Basin, Mediterranean, North, East, West, South-west, Centre-east and Centre-west.
2. Germany
   In Germany, interviews were carried out with 403 children at 34 sampling points which included four or more towns from each of the

following seven regions: North, West, South, North-east, Hessen/Saan/
Pfalz, Baden-Württemberg and Thuringen/Saxony.
3. Italy
   In Italy, interviews were carried out with 408 children at 25 sampling
   points which included four or more towns from each of five regions:
   North-east, North-west, South, Centre and Islands.
4. UK
   In the UK, interviews were carried out with 402 children at 25 sampling
   points, which included one or more locations in each of 11 regions:
   Scotland, North, Yorkshire and Humberside, North-west, Midlands,
   Greater London, Outer Metropolitan district, Outer South-east, Wales,
   South-west and East Anglia.

9.26 Standard questionnaire

An asterisk (*) indicates that multiple answers were allowed for that
question.

Introduction

I am from the Children's Research Unit and we are conducting a survey
amongst people aged 8–15. The survey is about food. First, may I check
some details about you?

(Enter CLASSIFICATION DATA)

- QA1*   I want to talk to you about food and nutrition, but first of all,
          can you tell me what you had to eat and drink for:
          - breakfast (today);
          - lunch (today);
          - dinner/supper/evening meal (yesterday).
- QA2*   What does the word nutrition mean to you? (PROBE
          Anything else?) (Explain if necessary)
- QA3*   Now, I have a list of foods here and I want you to show me
          which ones you think are Good for you/OK for you/Not so
          good for you. Foods and drinks were mixed between QA3,
          QA4 and QA5. (Note that in Italy, the 3-point scale was
          amended to: 'Very good for your health'; 'Good for your
          health'; 'Not so good for your health'.)

QA3	QA4	QA5
Bread	Butter	Chicken
Water	Biscuits	Sweets
Breakfast cereal	Sugar	French fries

Vegetables	Chocolate	Soup
Fruit	Pasta	Cakes
Meat	Margarine	Pizza
Fish	Crisps	Beer
Eggs	Ice cream	Ketchup
Milk	Fizzy drinks	Wine
Cheese	Burgers	Cider
		Yoghurt (France and Italy only)
		Olive oil (Italy only)

- QA6*   Where do you learn about nutrition?
  - Family
  - Friends
  - School teacher
  - Television
  - School
  - Magazines
  - Advertising
  - Other (specify)
- QA7   Do you think it is important that children learn about nutrition? If YES, how would you like to see this done?
  - Cooking classes at school.
  - TV programmes.
  - Magazines.
  - Special information packs for use at school.
  - Other (specify).
- QA8*   Who usually decides what you eat for:
  - breakfast during week;
  - breakfast during weekend;
  - lunch during week;
  - lunch during weekend;
  - dinner/supper during week;
  - dinner/supper during weekend.

  - Father - I do - Mother - School - Other
- QA9   Which meals do you usually eat with your family? (Definition of family = at least 1 parent present)
  - breakfast during week;
  - breakfast during weekend;
  - lunch during week;
  - lunch during weekend;
  - dinner/supper during week;
  - dinner/supper during weekend;
  - other (specify).

- QA10/11 Can you tell me if you agree or disagree with these statements or if you don't know?
    - 'It is best to eat small amounts of different foods rather than a lot of the same food'
    - 'To stay healthy, you should eat less fat'
    - 'Milk is good for strong bones'
    - 'Foods like sweets and ice cream are OK to eat, but not all the time'
    - 'It is important to eat foods like whole grain bread and cereals'
    - 'Fast foods are OK to eat every day'
    - 'The food you eat affects your health as you grow up'
    - 'Ready-made meals are just as good for you as home-made meals'
    - 'Exercise is just as important as the food you eat for staying healthy'
    - 'Tinned fruit and vegetables are just as good for you as fresh fruit and vegetables'
    - 'Chocolate is OK to eat every day'
- QA12*   For your body to stay healthy which of these does your body need?
    - Protein
    - Fat
    - Minerals
    - Vitamins
    - Salt
    - Fibre
    - Pills
    - Calories
    - Calcium
    - Sugar
    - Starch
- QA13   How much do you think you know/don't know about nutrition? Can you show me which of these statements applies to you the most?
    - I know a lot about nutrition.
    - I know some things about nutrition.
    - I don't know much about nutrition.
    - I don't know anything about nutrition.
- QB1   What is Salmonella?
    - Bacteria, which, when found in food, make you ill.
    - Type of pink salmon.
    - A common child disease.
    - Don't know.

- QB2   I would now like to read you a number of statements. Please tell me if you agree or disagree with these statements or if you don't know.
  - 'It is important to keep the kitchen clean'
  - 'Fresh foods are safer than frozen or tinned foods'
  - 'You should always cover food before putting it back in the fridge'
  - 'You should always wash fruit and vegetables before eating them'
  - 'Foods can go "off" from being kept at the wrong temperature'
  - 'You don't always need to wash your hands before you eat'
  - 'It is safer to drink milk straight from the cow than the milk one buys in the store'
  - 'The food we eat today is safer than ever before'
  - 'Foods containing E numbers are bad for you'
- QB3*   How do you get food poisoning?
  - From eating too much.
  - From harmful bacteria in food.
  - From the additives in food.
  - From the pesticides that are used on crops.
  - Don't know.

QB4*   Which of these things do 'use-by' dates tell you?
  - You need to eat the food before that date.
  - It's OK to eat it two weeks after the date mentioned.
  - It's the date by which stores need to sell that food.
  - Don't know.

10 International influences on children's food and drink

V. N. BALASUBRAMANYAM

10.1 Introduction

Food habits acquired during one's childhood die hard. Immigrants from India living in Britain may get accustomed to a British diet, develop a taste for fish and chips, but will nonetheless yearn for the peppery foods of their homeland. Indeed, one's accent, regarded as a sure give-away of one's origins, may be easier to lose than one's food habits acquired during childhood.

How are food and drink habits formed during childhood? Has the growth in the economic interdependence of nations influenced the food and drink habits of children? Do such influences, if any, promote cultural awareness and educate children in the ways and mores of life elsewhere than that in their homeland? This chapter addresses some of these issues. The discussion relates to food habits of children of all ages, including teenagers.

10.2 The influences on children's food and drink habits

10.2.1 Early childhood

A variety of factors and sources influence children's food and drink habits. In early childhood it is the habits of parents and family which influence children's food and drink habits. Once children reach school age they become exposed to external influences from a variety of sources. These may include friends at school, teachers, children's books, television, the food and drink of the countries visited on holidays with parents and the information acquired on weekly visits to the supermarket with parents.

10.2.2 Geography

These several influences are, of course, conditioned by the geography, religion, the ethnic composition of the population and the economic characteristics of the country in which the child and parents reside. These are also the principal factors which make up the culture of the country, broadly defined. Geography influences food and drink habits through its influence on the agriculture of the region or, more specifically, the type of foods the region produces. Although several examples of the influence of geography on food habits readily suggest themselves, the obvious one is the predominance of rice and rice-based foods in the diet of those living in the south of India and wheat and wheat-based foods in the diet of people from the north of India.

10.2.3   Religion

The influence of religion and religious edicts on food and drink habits may not be as obvious; several of these are accepted as the norm and the religious sanctions underlying them and their rationale are lost in the mists of time. Hindus do not eat beef, Muslims and Jews do not eat pork, practising Catholics stick to fish on Fridays and the Jain sect in India abhor meat, fowl and fish of any kind.

10.2.4   Ethnic composition

The ethnic composition of the population, as well as the history of a country, also influence the food habits of the people. There is, thus, a strong Spanish and Italian influence on the food habits of Americans – chilli con carne, enchiladas and pizza readily suggest themselves as examples of ethnic influences on the food habits of Americans. Indeed, in recent years the British, not well known for their cuisine or their adventurous food habits, have experienced a veritable gastronomic invasion with the growth of the immigrant population from the sub-continent and people of Indian origin from East Africa.

10.2.5   Economics

All of these influences have their international dimensions, particularly the influence of history and the ethnic origins of the people of a country. It is, however, the structure of the economy of a country which imparts a significant international dimension to the food and drink taste patterns of its people. Indeed, economic forces often override the influence of geography on the taste patterns of the inhabitants of a country. Singapore, for instance, has no domestic sources of food and is dependent entirely on imports. Admittedly the ethnic composition of Singapore's population, consisting of Chinese and Indians, influences the types of food available in the country. However, Singapore provides a sumptuous feast of varieties of foods, as any traveller to the island knows. This is because of the heavy orientation of Singapore to international trade and international investment.

10.2.6   *International trade and travel*

Indeed, international trade, international investment and international travel have all, in a manner of speaking, liberated consumers the world over from the strict diet specific to the region in which they reside. While international travel exposes one to new varieties of food and drink, international trade and international investment make available these new foods and, to an extent, also educate the consumer by creating an awareness of these new varieties of food and drink. In recent years, British citizens have increasingly ventured abroad for their holidays. In addition to the traditional European destinations, such as Spain and Greece, they are now venturing further afield into Asia and America. Both the innovative products the travel industry has produced, such as package holidays, and the decline in the real cost of travel have fuelled this growth in international travel. Children, accompanying their parents, appear to be even more adventurous in trying out the food and drink these foreign parts have to offer. Whether it be moussaka in Greece or paella in Spain or rogan josh in India, they do not hesitate to try them.

This introduction to foreign foods initiated by travel is fostered and developed by international trade and investment. The formation of the European Union (EU) has facilitated both the movement of people and products between member countries. It is not only the removal of artificial barriers to trade, such as tariffs and quotas, but also the business travel the EU has generated which accounts for the observed growth in trade between member countries. Regional groupings such as the EU and the much looser groupings such as the ASEAN are not the only reasons for the growth in world trade. The revolution in transportation and communications and a general move towards freer trade on the part of most trading nations are also significant factors in the rapid growth of world trade. Food and drink products figure prominently in this growth of world trade in recent years.

A significant characteristic of world trade in food and drink is worth noting. In the case of the principal trading nations of the world, including the UK, much of this trade consists of intra-industry trade as opposed to inter-industry trade (Greenaway and Milner, 1986). The former refers to trade in similar but differentiated products, for example, exchange of Cadbury's chocolates for the Swiss variety or the exchange of Guinness for Danish beer. In essence, intra-industry trade refers to trade in differentiated products which belong to an identifiable class. Inter-industry trade, on the other hand, refers to the exchange of products which are dissimilar, such as the exchange of tea for chemicals. In the case of the UK intra-industry trade in food as a group accounts for more than 60% of total trade. In the case of drinks, intra-industry trade is as high as 70%.

The growth in intra-industry trade is both a cause and consequence of converging taste patterns for food and drink in the world economy, principally so in the EU. It is also a consequence of growth in specialization in production, with the products embodying the producing country's socio-economic characteristics.

10.2.7   International investment

International investment has also contributed to the convergence of taste patterns in food and drink. It has done as much, if not more, than international trade in educating the consumer and in providing him/her with varieties of food and drink not available before. In fact, international investment rather than international trade is the preferred mode of international market penetration for most food and drink products because of their bulk and perishable nature. The UK is not only one of the leading overseas investors in the food and drink industry, but is also host to major international food and drink companies from abroad. The food and drink sector accounts for nearly a third of Britain's total stock of foreign direct investment valued at £247 billion in 1993. As a consequence of this two-way flow of investment, we find British firms producing one variety of food abroad, while foreign firms produce a similar product but with the emphasis on varieties which are not indigenous to British firms. Cadbury producing chocolates of a particular variety abroad and Mars and Nestlé producing another variety or varieties of chocolates in Britain, is a case in point. Other examples abound in the drinks industry group and in the breakfast cereals market.

10.2.8   Retailers

Two other sources of influence on food and drink taste patterns, both allied to international trade and investment, are noteworthy. First is the influence of the food and drink retailers. These include not only the major supermarket chains such as Sainsbury, ASDA and Marks and Spencer, but also a number of small specialist food and drink retailers. Second is the growing presence of restaurants serving a variety of ethnic and country-specific foods. In recent years a ubiquitous presence of Indian, Chinese and Thai restaurants has become evident in Britain. None of these are specific to Britain. Both the retail outlets selling a variety of foods from the world over and restaurants serving a variety of ethnic foods are to be found in most developed countries.

The major retail outlets contribute to the awareness of food and drink from other lands in many ways. First, they provide retail outlets for major manufacturers of these foods, both at home and abroad. Second, they provide the ingredients, often exotic ones, for preparation of foods from other lands. This they provide both to households and restaurants. Third, they advertise the availability of food and drink from a variety of countries both through their shelf displays and in the media. In addition to the retailers, food and drink manufacturers also advertise their wares. In fact, the incidence of advertising is especially marked in the case of the food and drink sector. The innumerable varieties and brands of food and drink produced compels the industry to engage in heavy advertising.

10.3 Peer pressure and its origins

The foregoing has sketched the principal factors which influence the availability and consumer awareness of food and drink from the world over. The discussion, however, has glossed over two issues. First, is it the case that producers of food and drink merely respond to consumer demand and taste patterns with advertising as a method of informing the consumer of their efforts in meeting his/her demand, or do the producers actively create wants and use advertising as a means of persuading the consumer to purchase the varieties of food and drink they produce? This is an age-old debate amongst economists which has yet to reach a settled conclusion. I would argue that advertising performs an essential role in informing and, in many cases, educating the consumer. Brands and trademarks are a method of ensuring quality and conveying information to the consumer, including the children amongst them (Balasubramanyam, 1994). However, this chapter is not the place to engage in this debate.

The most significant influences on children's food and drink habits are peer pressure and the preferences and habits of parents. Who are the peers and what factors influence their habits? The peers come from the upper income groups of the population – the ones who can afford international travel and exotic foods which are relatively expensive at the initial stage of their introduction to the markets. The consumption patterns of children from the upper income groups at school and outside at parties and picnics shape the habits of other children. The consumption of the upper income groups may be of the Veblenesque conspicuous consumption variety, driven by current fads and fashions. However, once such a pattern catches, it is only a matter of time before the food industry and the retailers move in and supply the foods at a price affordable to the average consumer.

How do the peers amongst the children acquire their food habits? As mentioned earlier, they acquire them from two sources – the family and, to a much lesser extent, from exposure to advertisements. The influence of peers on children's food habits is illustrated by the experience of a colleague – an immigrant from Nigeria. Muslims by birth and upbringing, he and his wife are not partial to sausages, but his seven-year-old son insists on having them for breakfast. The father satiates his son's appetite for sausages by driving sixty miles to the nearest butcher who can supply him with *halal* sausages. The child has acquired a taste for sausages having seen his friends at school eating them.

Most children from upper income group families travel abroad on holidays and return with a taste for foreign foods. These days the demand for foods of other countries tends to be substantial, resulting in their importation or production by the leading food firms. This sort of demonstration effect appears to be the major influence on children's food and drink habits.

Such a demonstration effect is reinforced by the food habits of the children of ethnic minorities. The taste for spicy foods from the sub-continent acquired by teenagers is mostly due to the presence of children of immigrants from India and Pakistan in local schools. The ubiquitous presence of restaurants serving ethnic foods further reinforces the effects of the school-based demonstration effect. In addition, clubs and disco-theques featuring Indian pop music groups also provide Indian food to complete the ambience of the sub-continent.

Children of immigrants born and brought up in Britain add another dimension to the spread of a taste for Indian food and drink. These children can be considered cultural amphibians, capable of surviving in both the Indian culture at home and the British mode of life at school and play. The teenagers in the ethnic groups are the ones who build a bridge between the two cultures, and it is their tastes in food and drink which are accepted and imitated. These are also the children who are much travelled between the sub-continent and Britain. It is revealing that they not only influence the taste patterns of the youth of Britain, but quite often in matters of food and dress they are also the trendsetters for the young in India. In this context, it is of interest that music recording companies based in Birmingham export Indian popular music from Britain to India.

10.4   Promoting cultural awareness: Britain and India

Have all these influences on the food habits of children promoted an awareness of the culture of other lands? Cultural awareness should be broadly defined here to include:

- knowledge of the ways and mores of life of people of other lands and climes;
- a broad understanding of their religion and beliefs; and
- a working knowledge of the history and geography of foreign lands.

For purposes of analysing this issue I rely upon Britain and India as case studies.

10.4.1   The Raj

The taste patterns of Indians have influenced English food and drink habits for a considerable length of time. Some of these habits date back to the days of the Raj. Thus, Indian imports, suitably Anglicised should be very familiar, such as mulligatawny soup, cold curried mutton and chutneys. These are the foods of the Raj, characteristic of old India.

It is, however, doubtful if these dishes formed the diet of English children in India during the days of the Raj. According to most accounts, the Mem Sahibs were keen on protecting their children from native influences, and they were duly packed off to England for schooling. Indian food, however, formed very much a part of the cuisine of the English civil servant and the soldier in India. It is a curious fact that while Indian food did make its way to the tables of the English servants of the Raj, English cuisine did not find much favour with the Indians of the day, not even the upper class Indians. No doubt Scotch whisky, Vimto, Ovaltine, Horlicks and Huntley and Palmers biscuits are, to this day, household names in India. But the influence of English food and drink habits on the Indians appear to be limited to alcoholic drinks and beverages. This cannot be attributed entirely to the bland nature of English food – it is presumably because Indian children rarely come into contact with English children, let alone English food. Children of the British servants of the Raj were either sent to exclusive schools in the hill stations of Ooty, Simla and Darjeeling where the English way of life (including food habits) was preserved, or they were sent back to England for schooling. Thus, while many an Indian child was familiar with English nursery rhymes, it had little contact with English children and their food habits. They seem to have remained impervious to English food habits for the rest of their lives; a fact which illustrates the thesis that if food habits are to persist, they have to be acquired during one's childhood, defined to include the teenage years.

10.4.2   The present day

Indian restaurants, cuisine from the sub-continent, the corner store of Mr Patel which sells Indian groceries, pickles and pappadums and is open all hours, including Sundays and Christmas Day, are all familiar sights in the Britain of today. In fact, they are a part of the cultural fabric of Britain. Children and teenagers of yester-years, with a taste for the foods of the sub-continent acquired during their school days, are now the principal customers of the Indian restaurants and shops dotted across Britain.

How has all this influenced cultural awareness? Those people that knew India in the old days, unfortunately a dying breed, were familiar with not only the food and drink of the sub-continent but also most aspects of the culture of India, simply because they lived and worked there. But what of the present generation who have acquired a taste for Indian food and drink through the demonstration effect of their peers, through travel and through the services of the major food and drink multinationals engaged in international trade and investment? They too have acquired a fund of knowledge about India and the customs, conventions and beliefs of her people. No child, teenager or adult would expect to see any beef dishes on the menus of Indian restaurants. This is because they know that the Indians revere and worship the cow, and most children will know of the reasons for the veneration of the cow by the Indians. The more curious of them may have dipped into the Indian epics, several English translations of which are available, and learned about Lord Krishna, one of the several Hindu gods known for his incarnation as a cowherd. This would no doubt raise their awareness of the history and religions of India.

10.4.3   Vegetarianism

Much more important is the influence of Indian food and food habits on the general attitudes towards food on the part of young British citizens. I have in mind here the spread of vegetarianism amongst the young, especially young girls. This is very much an Indian influence. The majority of Indians are not vegetarians, but the minority who are vegetarians account for a sizeable number in a country with a population of 900 million people. Amongst the immigrant population of Britain are to be found a substantial number of vegetarians, mostly those with origins in southern India and the state of Gujarat on the west coast. The way of life of their children, who are mostly vegetarians, has had a major influence on the food habits of children in Britain. More than this demonstration effect, the almost infinite variety of vegetarian and vegetable-based foods that Indian cuisine makes available is, in itself, a major reason for the spread of vegetarianism amongst the young. It is not only that Indian food consists of a variety of vegetarian dishes, but that Indians have demonstrated that vegetables and cereals can provide a balanced and appetizing diet; a lesson that many an English restaurant owner appears to have learned.

Along with the adoption of vegetarianism comes an awareness of the requirements of a balanced diet and the ingredients required to produce a balanced diet. If the health conscious young of today take a dim view of fast food chains, it is not only because of their awareness of the impact of such chains on the environment, but it is also because of their awareness of animal rights and the noxious influence of some meat-based foods on health. Such awareness is a consequence of the spread of vegetarianism, which is itself heavily influenced by the food habits of ethnic minorities in Britain.

Beyond promoting vegetarianism, international influences on food habits of the young have also promoted an awareness of the music and art of other lands. The manner in which food and drink are prepared, the manner in which they are served, the place where they are served and the company in which they are taken all add up to a cultural experience in itself. These are also factors which arouse the curiosity of the young to know more about the country and the people from which the food and drink originate. If such curiosity also influences the young to travel, it further promotes their education and knowledge of other countries and climes.

10.4.4   Advertising and brands

Lastly the economic characteristics of the food and drink industry, some of them specific to the two groups, also contribute to cultural awareness among the young. The specific characteristic I have in mind here is the advertising intensity of the industry and the heavy incidence of brands in the industry. The hackneyed phrase 'variety is the spice of life' is nowhere more true than in the case of the food and drink industry. Most companies, retailers and restaurants compete for markets on the basis of product differentiation encapsulated in brand names. The *Grocer* magazine, for instance, lists five major categories of cakes produced by Cadbury with more than sixty varieties and brands of cakes. Such differentiation and brands necessitate more advertising and market promotion than in most other industries.

Advertising and brands are a means of providing information about the characteristics of the product to the consumer. Such information often includes little nuggets relating to the origins of the product, the region from which it comes and, quite often, the brand name itself is suggestive of its origins. Many examples in this context relate to the drinks industry. Coca Cola, universally abbreviated to coke, is readily associated with the United States; many of the recent varieties of soft drinks invoke visions of the Caribbean; Tiger beer is readily recognized as a product of Singapore and Guinness is, of course, Irish to the core, even if brewed in England. The hero of *Crocodile Dundee* must have done more to promote the image of Australia with his promotion of Australian beers with their catchy brand names.

10.4.5   Developing countries

There is little doubt that food and drink from other lands have created an awareness of other cultures in Britain amongst the young. But has it done the same in developing countries. Rapidly growing and newly industrializing countries, including India and China, have opened their doors to foreign trade and foreign investment in the food and drink industries. Some of the leading food companies such as Cadbury have had a major presence in

India for quite some time. McDonalds, Coca Cola and Wimpey, however, are much more recent arrivals. It is the American fast food and drink firms, including the fast food chains and their indigenous versions, which appear to have exerted a strong influence on the food and drink habits of the young in India. Their influence is mostly confined to the affluent few. It is debatable, however, whether the spread of the food habits of the youth of America amongst the affluent young of India has served to arouse their cultural awareness or merely satisfy their yearning for westernization and what they perceive to be a sophisticated way of life.

10.5   Conclusion

This chapter is essentially, discursive. I have attempted to analyse the sources of international influence on the taste patterns and food habits of the young. Much of what has been discussed is born out of observation and is coloured by subjective interpretation of available facts. Few would dispute, however, the thesis that food and drink habits acquired during childhood stay for ever. In a society such as Britain which, because of its geography and colourful past, is not only exposed to the food and drink of other cultures and climes but also provides a home for a number of ethnic groups, children's food and drink habits reflect the influence of the cultural milieu they live in. Food and drink which, in many ways, make a major contribution to our understanding of the ways and mores of life of countries other than the one into which we are born.

Tolerance and appreciation of other ways of life which cultural awareness promotes are essential for peace and harmony in a multi-racial society. To the extent that food and drink promote cultural awareness, they contribute to the much sought after peace and harmony.

References

Balasubramanyam, V.N. and Salisu, M.A. (1994), Brands and the Alcoholic Drinks Industry. In *Adding Value: Brands and Marketing In Food and Drink*, Jones, G. and Morgan, N.J. (Eds), Routledge, London.
Greenaway, D. and Milner, C. (1986), *The Economics of Intra-industry Trade*, Blackwell, Oxford.

Index